THE IMPACT OF COVID-19
ON THE CAREERS OF WOMEN IN ACADEMIC
SCIENCES, ENGINEERING, AND MEDICINE

Eve Higginbotham and Maria Lund Dahlberg, *Editors*

Committee on Investigating the Potential Impacts of COVID-19 on the Careers of Women in Academic Science, Engineering, and Medicine

Committee on Women in Science, Engineering, and Medicine

Policy and Global Affairs

A Consensus Study Report of

The National Academies of
SCIENCES · ENGINEERING · MEDICINE

THE NATIONAL ACADEMIES PRESS
Washington, DC
www.nap.edu

THE NATIONAL ACADEMIES PRESS   500 Fifth Street, NW   Washington, DC 20001

This activity was supported by contracts between the National Institutes of Health (IDIQ contract HHSN263201800029I, TO #75N98020F00008), the National Science Foundation (Award #OIA-1762395), the National Institute of Standards and Technology (Contract #SB134117CQ0017, TO #1333ND20FNB100250), the Alfred P. Sloan Foundation (Award #G-2020-13993), and the Doris Duke Charitable Foundation (Award #2020136). Any opinions, findings, conclusions, or recommendations expressed in this publication do not necessarily reflect the views of any organization or agency that provided support for the project.

International Standard Book Number-13: 978-0-309-26837-0
International Standard Book Number-10: 0-309-26837-0
Digital Object Identifier: https://doi.org/10.17226/26061
Library of Congress Control Number: 2021937008

Additional copies of this publication are available from the National Academies Press, 500 Fifth Street, NW, Keck 360, Washington, DC 20001; (800) 624-6242 or (202) 334-3313; http://www.nap.edu.

Copyright 2021 by the National Academy of Sciences. All rights reserved.

Printed in the United States of America

Suggested citation: National Academies of Sciences, Engineering, and Medicine. 2021. *The Impact of COVID-19 on the Careers of Women in Academic Sciences, Engineering, and Medicine*. Washington, DC: The National Academies Press. https://doi.org/10.17226/26061.

The National Academies of
SCIENCES · ENGINEERING · MEDICINE

The **National Academy of Sciences** was established in 1863 by an Act of Congress, signed by President Lincoln, as a private, nongovernmental institution to advise the nation on issues related to science and technology. Members are elected by their peers for outstanding contributions to research. Dr. Marcia McNutt is president.

The **National Academy of Engineering** was established in 1964 under the charter of the National Academy of Sciences to bring the practices of engineering to advising the nation. Members are elected by their peers for extraordinary contributions to engineering. Dr. John L. Anderson is president.

The **National Academy of Medicine** (formerly the Institute of Medicine) was established in 1970 under the charter of the National Academy of Sciences to advise the nation on medical and health issues. Members are elected by their peers for distinguished contributions to medicine and health. Dr. Victor J. Dzau is president.

The three Academies work together as the **National Academies of Sciences, Engineering, and Medicine** to provide independent, objective analysis and advice to the nation and conduct other activities to solve complex problems and inform public policy decisions. The National Academies also encourage education and research, recognize outstanding contributions to knowledge, and increase public understanding in matters of science, engineering, and medicine.

Learn more about the National Academies of Sciences, Engineering, and Medicine at www.nationalacademies.org.

The National Academies of
SCIENCES • ENGINEERING • MEDICINE

Consensus Study Reports published by the National Academies of Sciences, Engineering, and Medicine document the evidence-based consensus on the study's statement of task by an authoring committee of experts. Reports typically include findings, conclusions, and recommendations based on information gathered by the committee and the committee's deliberations. Each report has been subjected to a rigorous and independent peer-review process and it represents the position of the National Academies on the statement of task.

Proceedings published by the National Academies of Sciences, Engineering, and Medicine chronicle the presentations and discussions at a workshop, symposium, or other event convened by the National Academies. The statements and opinions contained in proceedings are those of the participants and are not endorsed by other participants, the planning committee, or the National Academies.

For information about other products and activities of the National Academies, please visit www.nationalacademies.org/about/whatwedo.

COMMITTEE ON THE INVESTIGATING THE POTENTIAL IMPACTS OF COVID-19 ON THE CAREERS OF WOMEN IN ACADEMIC SCIENCE, ENGINEERING, AND MEDICINE

EVE J. HIGGINBOTHAM (NAM) (*Chair*), Vice Dean of Inclusion and Diversity, Senior Fellow in the Leonard Davis Institute of Health, and Professor of Ophthalmology, University of Pennsylvania

ELENA FUENTES-AFFLICK (NAM), Professor and Vice Chair of Pediatrics, Chief of Pediatrics at the Zuckerberg San Francisco General Hospital, and Vice Dean for Academic Affairs in the School of Medicine, University of California, San Francisco

LESLIE D. GONZALES, Associate Professor in the Higher, Adult, and Lifelong Learning Unit in the College of Education, Michigan State University

JENI HART, Dean of the Graduate School and Vice Provost for Graduate Studies and Professor of Higher Education in the Department of Educational Leadership and Policy Analysis, University of Missouri

RESHMA JAGSI, Newman Family Professor and Deputy Chair in the Department of Radiation Oncology and Director of the Center for Bioethics and Social Sciences in Medicine, University of Michigan

LEAH JAMIESON (NAE), Ransburg Distinguished Professor of Electrical and Computer Engineering, John A. Edwardson Dean Emerita of the College of Engineering, and Professor by courtesy in the School of Engineering Education, Purdue University

ERICK C. JONES, George and Elizabeth Pickett Endowed Professor and Associate Dean for Graduate Studies in the College of Engineering, University of Texas at Arlington

BERONDA MONTGOMERY, Professor of Biochemistry and Molecular Biology and of Microbiology and Molecular Genetics in the Department of Energy Plant Research Laboratory, and Interim Assistant Vice President of Research and Innovation, Michigan State University

KYLE MYERS, Assistant Professor of Business Administration in the Technology and Operations Management Unit, Harvard Business School

RENETTA TULL, Vice Chancellor of Diversity, Equity and Inclusion, University of California, Davis

Study Staff

MARIA LUND DAHLBERG, Study Director
ARIELLE BAKER, Program Officer
IMANI BRAXTON-ALLEN, Senior Program Assistant
JEENA M. THOMAS, Program Officer
THOMAS RUDIN, Director, Committee on Women in Science, Engineering, and Medicine (*until November 2020*)
BARDIA MASSOUDKHAN, Senior Finance Business Partner
JOE ALPER, Consulting Editor

COMMITTEE ON WOMEN IN SCIENCE, ENGINEERING, AND MEDICINE

GILDA BARABINO (NAE/NAM) (*Current Chair*), President, Olin College of Engineering
JOAN WENNSTROM BENNETT (NAS) (*Previous Chair*), Distinguished Professor of Plant Biology and Pathology and Associate Vice President in the Office for Promotion of Women in Science, Engineering, and Mathematics, Rutgers University (*Chair until December 2020*)
NANCY ANDREWS, (NAS/NAM) Dean of the Duke University School of Medicine and Vice Chancellor for Academic Affairs, Nanaline H. Duke Professor of Pediatrics, and Professor of Pharmacology and Cancer Biology, Duke University
MAY BERENBAUM, (NAS) Professor and Head of Entomology, University of Illinois at Urbana-Champaign
ANA MARI CAUCE, President, University of Washington (*until December 2020*)
VALERIE CONN, President, Science Philanthropy Alliance
MACHI DILWORTH, Vice President (Retired), Gender Equality and Human Resource Development, Okinawa Institute of Science and Technology Graduate University (*until December 2020*)
EVELYNN M. HAMMONDS, (NAM) Barbara Gutmann Rosenkrantz Professor of the History of Science, Professor of African and African American Studies, and Chair of the Department of the History of Science, Harvard University
HILARY LAPPIN-SCOTT, Professor, Cardiff University, United Kingdom
ED LAZOWSKA, (NAE) Bill & Melinda Gates Chair, Paul G. Allen School of Computer Science & Engineering, University of Washington
VALERIE TAYLOR, Director, Mathematics and Computer Science Division, U.S. Department of Energy's Argonne National Laboratory

Ex-Officio Members

ELENA FUENTES-AFFLICK, (NAM) Home Secretary of the National Academy of Medicine, Professor and Vice Chair of Pediatrics, Chief of Pediatrics at the Zuckerberg San Francisco General Hospital, and Vice Dean for Academic Affairs in the School of Medicine, University of California, San Francisco
CAROL K. HALL, (NAE) Home Secretary of the National Academy of Engineering, Camille Dreyfus Distinguished University Professor in the Department of Chemical and Biomolecular Engineering, North Carolina State University

SUSAN R. WESSLER, (NAS) Home Secretary of the National Academy of Sciences, Neil A. and Rochelle A. Campbell Presidential Chair for Innovation in Science Education, University of California, Riverside

Committee Staff

AUSTEN APPLEGATE, Research Associate
ARIELLE BAKER, Program Officer
ASHLEY BEAR, Senior Program Officer
FRAZIER BENYA, Senior Program Officer
IMANI BRAXTON-ALLEN, Senior Program Assistant
MARIA LUND DAHLBERG, Senior Program Officer
MARIE HARTON, Program Officer
ALEX HELMAN, Program Officer
REBEKAH HUTTON, Program Officer
THOMAS RUDIN, Board Director (*until November 2020*)
JEENA M. THOMAS, Program Officer
JOHN VERAS, Senior Program Assistant
MARQUITA WHITING, Senior Program Assistant

Preface

"A year like no other"[1] is an often-repeated phrase, given our collective experiences during 2020. The day before the start of the New Year, a mysterious illness was reported in China after dozens of people visited a live animal market in Wuhan. The first death was reported in mid-January. By the end of January, the World Health Organization declared a global health emergency, and the first suspected case was reported in the United States at the end of February. It was not until March 2020 that a federal emergency was declared, and since then we have been on what feels like a collective roller coaster, punctuated by fear, sadness, and hope.[2]

As I write this Preface during the close of 2020, the Food and Drug Administration provided emergency approval of a COVID-19 vaccine, the United States once again experienced its highest daily reported number of COVID-19–related deaths, the Supreme Court recently dismissed a lawsuit intended to overturn the outcome of last month's presidential race, and institutions and organizations across the nation are shifting to more intentional strategies to address structural racism. Added to the backdrop of this theater of disruption, there have been record-breaking fires on the West Coast and hurricanes and tornados elsewhere. Once again, the significant rise in COVID-19 cases across the country has translated into school districts, restaurants, and brick-and-mortar businesses closing down. To complete the image of this moment, the passage of a much-needed federal relief bill remains uncertain. Even as I reflect on the ongoing effects of

---

[1] Pixabay/CC0 Public Domain. COVID-19: Twelve key milestones in a year like no other. November 26, 2020. https://medicalxpress.com/news/2020-11-covid-twelve-key-milestones-year.html.

[2] E. Schumaker. Timeline: How coronavirus got started. ABCNews. September 22, 2020, 11:55 a.m. Available at https://abcnews.go.com/Health/timeline-coronavirus-started/story?id=69435165.

climate change, protests calling for social justice, and the impact of economic volatility, it is clear that this year is only a preview of similar confluences of disruptors to come.

It is incredible to believe that less than 1 year ago, the report *Promising Practices for Addressing the Underrepresentation of Women in Science, Engineering, and Medicine*[3] (the *Promising Practices* report) was released by the National Academies of Sciences, Engineering, and Medicine. As a reviewer of that consensus study report, I appreciated the summary of the current state of the representation of women across disciplines, the focus on intersectionality, particularly Women of Color, and the impressive list of evidence-based interventions that was advanced as promising practices. It provided a platform for launching, with renewed vigor, initiatives that may enable us to turn the corner and accelerate the advancement of women in academia. Considering that more than half of the population in the United States identifies as women, if we do not address our underrepresentation in science, technology, engineering, mathematics, and medicine (STEMM), the country is leaving an enormous magnitude of intellectual capital on the table. Innovation, enhanced decision-making, and profitability of corporations have been attributed to greater diversity.[4] Indeed, given the enormous value of diversity, the subtitle of the *Promising Practices* report, "Opening Doors," inspires an element of optimism and hope that continued progress is within our grasp.

It is fortunate that the sponsors of this current report, the National Institutes of Health, Alfred P. Sloan Foundation, Doris Duke Charitable Foundation, National Science Foundation, and the National Institute of Standards and Technology, partnered with the National Academies to stand up this committee to investigate the impact of COVID-19 on the academic careers of women in STEMM. After all, in these unprecedented times, the interventions enumerated in the *Promising Practices* report were no longer grounded in an environment that was encased in certainty. The questions that were embedded in our Statement of Task for this report were relevant and critically important to the future representation and viability of academic women in STEMM. Was there harm imposed on women when specific interventions were undertaken? Are women and men affected differently when these interventions are implemented? What are the unique challenges that women are facing during this COVID-19 pandemic? What are the early indicators of impacts to the career trajectories of women in STEMM? These are critical questions to better understand how we as an academic community can emerge from this pandemic with continued advancement of women rather than a

---

[3] National Academies of Sciences, Engineering, and Medicine. 2020. *Promising Practices for Addressing the Underrepresentation of Women in Science, Engineering, and Medicine: Opening Doors.* Washington, DC: The National Academies Press. https://doi.org/10.17226/25585.

[4] U.S. Glass Ceiling Commission. 1995. Good for Business: Making Full Use of the Nation's Human Capital. Washington, DC: U.S. Government Printing Office. http://digitalcommons.ilr.cornell.edu/key_workplace/ 116/.

reduction in the engagement of women in our scientific workforce and a deflation in the expanse of dreams of durable professional careers. An appropriate subtitle for this report, building on the work of the *Promising Practices* report, may have been "Keeping the Doors Open—Key Questions Unanswered."

Predictably, there was not a massive body of peer-reviewed literature upon which the committee could determine its list of findings and proposed research questions. However, five commissioned papers in key topical areas provided a rich resource for the committee's deliberations. Career trajectories, work-life integration, collaboration and networking, leadership and decision-making, and mental health and well-being made up our foci. We also addressed important cross-cutting issues such as the impact of the pandemic on Women of Color. From a range of literature sources, including reviews of relevant material from before the COVID-19 pandemic, the authors of the commissioned papers provided the committee with evidence to support a set of key findings and develop research questions for further investigation. Given how much is at stake, understanding the impact of COVID-19 on the careers of women in STEMM is of utmost importance.

On a personal level, I relate to many of the challenges described in this report and admit that the intensification of them during the current pandemic is nearly unimaginable. My caregiving responsibilities and the difficulties of integrating my personal and professional responsibilities were heightened in the middle of my career, following Hurricane Katrina. My parents, who were in their 90s at the time, came to live with my husband and me in Maryland. Although our family home in New Orleans was fine, I did not believe the health-care system and social services infrastructure in my hometown had recovered enough to effectively care for my parents living alone. My husband and I quickly shifted to preparing regular meals at home, hurriedly sought out caregivers to come in periodically to assist us, and ensured that we made all of their necessary doctors' appointments between our professional commitments. Taking up caregiving responsibilities after work, preparing the pill boxes late at night, and canceling visiting professorships at the last minute because of periodic emergencies and hospital admissions were new challenges that I had not previously experienced in my adult working life. Indeed, I wrote fewer papers and accepted fewer speaking invitations, deepening my personal concern that my profession would perceive these absences more harshly, considering that I am a woman, and particularly as a Woman of Color.

I recognize that if I had experienced these challenges earlier in my career, I would never have made it to positions of department chair or dean. With the added constraints and circumstances of the COVID-19 pandemic, I acknowledge that mothers today have taken on the roles of teacher and caregiver, often without the possibility of external assistance that was available to me. Moreover, I was fortunate to have a supportive spouse; however, in the absence of a partner, there is rarely time for self-care, scholarship, or other professional responsibilities. Our gendered roles in society take on added emphasis during periods of stress, the timing of which can make or break careers. It is time for institutions to consider new strategies to support the careers of women across their entire career timeline and embrace professional norms to alternative constructs other than a "hustle" culture.

It is my hope that this report will not only advance the discussion about how best to enhance the representation and vitality of academic women in STEMM, but create an awareness about the adverse impact that these unprecedented times will have on women going forward. It is also my hope that the research questions posed by this committee will translate into enduring solutions that will strengthen institutional interventions to weather future disruptions.

I appreciate the wisdom of the National Academies and our sponsors to take on this topic, the trusted guidance of the staff of the National Academies, the authors of our commissioned papers who sought the evidence that was needed, and the generosity of valuable time and energy of my fellow committee members who were so willing to expertly share their wisdom on prolonged Zoom calls. At our last meeting, I closed our session with the following quote, which has been attributed to Dr. Martin Luther King Jr.: "The arc of the moral universe is long but it bends toward justice."[5] It is my dream that the promise of the research that evolves from the questions that have been posed will not only shed a light on the opportunities to listen, learn, strategize, and implement, but will effectively facilitate our shared journey toward gender equity in academia.

Eve Higginbotham, *Chair*
Committee on Investigating the Potential the Impacts of COVID-19 on the Careers of Women in Academic Science, Engineering, and Medicine

---

[5] Dr. Martin Luther King Jr., quoted by the National Museum of African American History and Culture. https://www.si.edu/spotlight/mlk?page=4&iframe=true.

Acknowledgment of Reviewers

This Consensus Study Report was reviewed in draft form by individuals chosen for their diverse perspectives and technical expertise. The purpose of this independent review is to provide candid and critical comments that will assist the National Academies of Sciences, Engineering, and Medicine in making each published report as sound as possible and to ensure that it meets the institutional standards for quality, objectivity, evidence, and responsiveness to the study charge. The review comments and draft manuscript remain confidential to protect the integrity of the deliberative process.

We thank the following individuals for their review of this report: Huda Akil, University of Michigan; Robin Bell, Columbia University; Michelle Cardel, University of Florida; Reginald DesRoches, Rice University; Kathryn Holland, University of Nebraska; Monica Johnson, Washington State University; Karen Kafadar, University of Virginia; Christy Lemak, University of Alabama; Eleni Linos, Stanford University; Tony Liss, The City College of New York; Nancy Rigotti, Massachusetts General Hospital; and Sonia Zárate, Howard Hughes Medical Institute.

Although the reviewers listed above provided many constructive comments and suggestions, they were not asked to endorse the conclusions or recommendations of this report nor did they see the final draft before its release. The review of this report was overseen by Katherine Freeman, Pennsylvania State University, and Bryna Kra, Northwestern University. They were responsible for making certain that an independent examination of this report was carried out in accordance with the standards of the National Academies and that all review comments were carefully considered. Responsibility for the final content rests entirely with the authoring committee and the National Academies.

Contents

Summary   1

1   INTRODUCTION   13
    Report Context, 14
    Beyond the COVID-19 Pandemic, 17
    Study Process, 20
    About the Report, 24

2   OCTOBER 2020 WOMEN IN STEMM FACULTY SURVEY
    ON WORK-LIFE EFFECTS OF THE COVID-19 PANDEMIC   27
    Introduction, 28
    Preferences and Changes in Number of Days Working at Home, 28
    Effects of the COVID-19 Pandemic on Work Productivity,
        Well-being, Childcare and Household Labor, and Eldercare, 30
    Coping Strategies for Blurred Boundaries and Domestic Labor, 38
    Actual versus Desired University Accommodations
        Post-COVID-19 Pandemic, 41
    Highlights from Non-Tenure-Track Faculty Responses, 42
    Conclusions, 45

3   ACADEMIC PRODUCTIVITY AND
    INSTITUTIONAL RESPONSES   47
    Introduction, 47
    Broader Labor Market Effects of the COVID-19 Pandemic, 48

Effects of the COVID-19 Pandemic on Academic Productivity in 2020, 49
Effects of Institutional Responses to the COVID-19 Pandemic on Academic Careers and Productivity, 54
Conclusions, 56

4 WORK-LIFE BOUNDARIES AND GENDERED DIVISIONS OF LABOR 57
Introduction, 57
Pre-COVID-19 Pandemic Work-Life Literature Overview, 58
Post-COVID-19 Pandemic Literature: Changes to Boundaries, Boundary Control, and Well-being, 66
Conclusions, 68

5 COLLABORATION, NETWORKS, AND ROLE OF PROFESSIONAL ORGANIZATIONS 69
Introduction, 69
Historical Events and the Impacts on Collaborations and Networking, 70
Effects of the COVID-19 Pandemic on Collaborations and Networking, 70
Effects of the COVID-19 Pandemic on Professional Organizations and Networks, 73
Effects of the COVID-19 Pandemic on Conferencing, 76
The Role of Mentorship and Sponsorship During the COVID-19 Pandemic, 79
Conclusion, 80

6 ACADEMIC LEADERSHIP AND DECISION-MAKING 83
Introduction, 83
Effects of the COVID-19 Pandemic on Women in Academic Leadership Positions, 85
COVID-19 Pandemic Decision-Making and Effects on Gender Inequalities, 85
Decision-Making During the COVID-19 Pandemic, 88
Leadership and Decision-Making to Address Crises and Inequities, 91
Data Gaps on Academic Leadership and Decision-Making, 94
Conclusion, 94

7 MENTAL HEALTH AND WELL-BEING 95
Introduction, 95
Effects of Isolation and Societal Stress for Women in STEMM, 96

Effects of COVID-19 Pandemic-Related Stress on Women
in STEMM, 97
The COVID-19 Pandemic and the Mental and Physical Health
of Women in STEMM, 99
Conclusion, 108

8 MAJOR FINDINGS AND RESEARCH QUESTIONS   109
   Introduction, 109
   Major Findings, 111
   Research Questions, 114

GLOSSARY   119

REFERENCES   125

APPENDIXES

Appendix A: Literature Review Terms and Survey
Methodology for "Boundaryless Work: The Impact of
COVID-19 on Work-Life Boundary Management, Integration,
and Gendered Divisions of Labor for Academic Women
in STEMM," by Ellen Ernst Kossek, Tammy D. Allen, and
Tracy L. Dumas   155

Appendix B: Methodology and Data Sources for the
"Academic STEMM Labor Market, Productivity, and
Institutional Responses," by Felicia A. Jefferson,
Matthew T. Hora, Sabrina L. Pickens, and Hal Salzman   161

Appendix C: Material Selection Process for "The Impact
of COVID-19 on Collaboration, Mentorship and Sponsorship,
and Role of Networks and Professional Organizations,"
by Misty Heggeness and Rochelle Williams   163

Appendix D: Committee Biographies   169

Boxes, Figures, and Tables

BOXES

Box 1-1	Statement of Task	20
Box 1-2	On the Use of Language	21
Box 5-1	Effects of the COVID-19 Pandemic on International and Distanced Collaborations	72
Box 6-1	Changing Nature of Decision-Making Under Academic Capitalism and the Gig Academy	86

FIGURES

Figure 2-1	Summary of effects of COVID-19 on the work effectiveness and productivity of women in academic STEMM from the October 2020 survey.	31
Figure 2-2	Challenges and coping strategies related to childcare demands reported in the October 2020 survey.	33
Figure 2-3	Challenges and coping strategies related to eldercare demands reported in the October 2020 survey.	37
Figure 2-4	Boundary management tactics and other coping strategies reported in the October 2020 survey.	39
Figure 4-1	Types of work-nonwork boundary management interruption styles.	60

TABLES

Table 7-1	Validated Measurements of Mental Well-being	104
Table 7-2	Risk and Resilience Factors: Documentation from Health-Care Workers Extrapolated to Academic Women in STEMM	107
Table A-1	Listservs that Posted the Anonymous Survey Link for the Work-Life Boundaries Paper	156
Table A-2	Sample Description for October 2020 Survey of Women in Academic STEMM Faculty	157
Table A-3	Topics for the October 2020 Survey	159
Table A-4	Search Terms and Numbers of Results for the Literature Review Conducted by Kossek, Dumas, and Allen	160
Table C-1	Professional Associations Reviewed by Misty Heggeness and Rochelle Williams	164

Summary

Spring 2020 changed how nearly everyone conducted their personal and professional lives, within science, technology, engineering, mathematics, and medicine (STEMM) and beyond. For academic STEMM, the disruptions caused by the COVID-19 pandemic ranged from delayed experiments in individual laboratories to transformed or canceled global scientific conferences. People shifted classes to virtual platforms and negotiated with family members for space in their homes from which to work. This changed reality blurred the boundaries between work and nonwork, infusing ambiguity into everyday activities. While adaptations that allowed people to connect became more common, the evidence available at the end of 2020 suggests that the disruptions caused by the COVID-19 pandemic endangered the engagement, experience, and retention of women in academic STEMM, and may roll back some of the achievement gains made by women in the academy to date.

By the end of 2020, it was well documented that the COVID-19 pandemic was particularly detrimental to vulnerable populations, such as People of Color and elderly individuals, and had a devastating effect on the economy, particularly brick-and-mortar retail and hospitality and food services. Less rigorously documented was how the COVID-19 pandemic added to and amplified gendered expectations for women in academic STEMM during its first several months. Although women shared firsthand accounts of new and enduring challenges, the overall effects of the COVID-19 pandemic were challenging to quantify using rigorous, data-driven methods less than 1 year after it began. To better understand and gather the available evidence on the potential impacts of the COVID-19 pandemic on the careers of women in academic STEMM, an ad hoc committee was appointed by the National Academies of Sciences, Engineering, and Medicine in late summer 2020.

Less than 6 months earlier, the National Academies released the consensus study report *Promising Practices for Addressing the Underrepresentation of Women in Science, Engineering, and Medicine: Opening Doors* (the *Promising Practices* report). While that report called attention to the challenges that women in STEMM experience and presented evidence-based recommendations to address the well-established structural barriers that impede the advancement of women in STEMM (NASEM, 2020), the actions it identified were not conceived within the context of a pandemic, an economic downturn, or the emergence of national protests against structural racism.

ABOUT THE REPORT

This report arose out of the need to expeditiously identify, name, and document how the COVID-19 pandemic disrupted the careers of women in academic STEMM during the initial 9-month period from March to December 2020, and to consider how these disruptions—both positive and negative—might shape future progress for women in academic STEMM. The committee's task was to build on the *Promising Practices* report and examine the COVID-19 pandemic's potential influences on women in academic STEMM. Preliminary evidence indicated that such disruptions could have both short- and long-term consequences, and will likely vary across institution type (e.g., community colleges, baccalaureate-granting institutions, doctoral-granting and research universities, Historically Black Colleges and Universities, Hispanic-Serving Institutions, and Tribal Colleges); career stage or focus (e.g., graduate student; postdoctoral scholar; medical resident; clinician; tenure-stream, tenured, full-time non-tenure-track, and adjunct faculty); academic rank (e.g., assistant professor, associate professor, full professor); and personal characteristics, including family structure, caregiving responsibilities, and behavioral health status. Developing a comprehensive understanding of the nuanced ways these disruptions have manifested may help the academic community emerge from the COVID-19 pandemic ready to mitigate any long-term negative consequences the COVID-19 pandemic might have on the continued advancement of women in the academic STEMM workforce. It may also help the academic community build on the adaptations and opportunities that have emerged during the course of the COVID-19 pandemic.

To inform its deliberations, findings, and research questions, the committee commissioned five papers. Each paper focused on a unique aspect of how the COVID-19 pandemic affected women in STEMM academics during 2020. The topics of the five papers as commissioned (and their authors) are as follows: the Impact of COVID-19 on (1) Tenure Clocks, the Evaluation of Productivity, and Academic STEMM Career Trajectories (Felicia A. Jefferson, Matthew T. Hora,

Sabrina L. Pickens, and Hal Salzman); (2) Boundary Management, Work-Life Integrations, and Household Labor (Ellen Kossek, Tammy D. Allen, and Tracy Dumas); (3) Collaboration, Mentorship and Sponsorship, and the Role of Networks and Professional Organizations (Rochelle Williams and Misty Heggeness); (4) Academic Leadership and Decision-Making (Adrianna Kezar); and (5) the Mental Health and Well-being of Women in STEMM (C. Neill Epperson, Elizabeth Harry, Judith G. Regensteiner, and Angie Ribera).

The central chapters of the report are based on the final drafts of these five papers. Each chapter provides key insights about how the COVID-19 pandemic had affected the careers of women in academic STEMM fields in 2020, approaching this core concept from different disciplinary perspectives. Chapter 2 sets the stage for the ensuing chapters and presents the results of a survey conducted in October 2020, providing a window into the very personal perspectives offered by respondents;[1] Chapters 3 through 7 review literature and concepts established before the COVID-19 pandemic, summarize the preliminary evidence and data on the impacts of the COVID-19 pandemic during 2020 from the perspective of that field, and—where possible—speculate about potential long-term implications. Taken together, these six different approaches form a single unified description of the potential impacts of the COVID-19 pandemic on the careers of women in STEMM during 2020.

BEFORE THE COVID-19 PANDEMIC

Advances in knowledge and practice in academic STEMM demand and benefit from a diversity of perspectives, including people who represent different genders, ethnicities, and ancestries. Different perspectives contribute unique "vectors of skills, experiences, and talents" to the STEMM enterprise (Page, 2007, 2008, 2019). However, women remain underrepresented in STEMM, with both societal and institutional inequities contributing both to this persistent underrepresentation and to the disproportionate burdens many women face in academic STEMM fields.

The organizational structures of colleges and universities, as well as the leadership and the decision-making context, are important determinants of gender equity. Women have a long history of underrepresentation in academic institutional leadership roles (Glazer-Raymo, 2001)—Women of Color even more so. Women also face numerous barriers, such as a "chilly climate," as well as organizational work policies that make it challenging for women to succeed, including tenure policies that make it challenging for women to have families (Maranto and Griffin, 2011; Sandler and Hall, 1986). Research shows that women are less likely to receive either mentoring or benefit from the sponsorship of senior academics,

---

[1] Details about the survey are provided in Chapter 2 and Appendix A. The survey was designed and fielded by Ellen Kossek, Tammy D. Allen, and Tracy Dumas.

have limited or no access to support structures, have less access to networks that would help them to move up in the ranks of administration, and become more isolated as they advance in the academic hierarchy (Maranto and Griffin, 2011; Sandler and Hall, 1986).

Institutional policies have often incentivized successful tenure-track faculty to secure substantial grants, patents, and licenses, while those same policies place a lower value on service work, mentoring, and student support—typically taken on by women and Faculty of Color—when it comes to deciding on tenure and promotion. However, nearly 70 percent of faculty are not on the tenure track, and women are overrepresented in this group (AAUP, 2020a).

During stressful times, those who are systemically disadvantaged are more likely to experience additional strain and instability than those who have an established reputation, a stable salary commitment, and power. Women of Color are impacted more significantly than others, given the layering of gender-bias and racism contributing to their career trajectories. In STEMM, desirable attributes are generally granted to those who adhere to masculine and majority norms. Women in academic STEMM are more likely to be early in their career, have a lower salary regardless of professional ranking in STEMM, be a single parent or a primary caregiver, and report experiencing greater work-related stress and discrimination in the workplace or their community. In addition, the caregiving responsibilities that often fall on the shoulders of women cuts across career timeline and rank (e.g., graduate student, postdoctoral scholar, non-tenure-track and other contingent faculty, tenure-track faculty), institution type, and scientific discipline.

IMPACT OF THE COVID-19 PANDEMIC

The preliminary evidence gathered in this report indicates that the COVID-19 pandemic has negatively affected the productivity, boundary setting and boundary control, networking and community building, and mental well-being of women in academic STEMM. It has led to school closures, shifting caregiving responsibilities onto parents and guardians, which has disproportionately negative outcomes for women across all sectors. Within STEMM, collaborations have been disrupted, career progressions have been paused, and women are facing challenges associated with gendered effects of remote work conflicting with caregiving responsibilities.

Because women were underrepresented across most STEMM fields, particularly in the upper echelons, women are more likely to experience academic isolation, including limited access to mentors, sponsors, and role models that share gender, racial, or ethnic identities. Coupled with the physical isolation stipulated by public health responses to the COVID-19 pandemic, women in academic STEMM have been isolated within their fields, networks, and communities, putting at risk the progress they have made in building networks and maintaining collaborations.[2]

Furthermore, women working in STEMM disciplines have begun to experience additional disruptions that may affect their academic productivity and careers. Preliminary evidence from 2020 suggests that the COVID-19 pandemic affected women's ability to engage actively in collaborations. For women in STEMM with children or other dependent care responsibilities, many had significantly less time in the day to network and engage in collaborations because of increased nonwork tasks (Heggeness, 2020; Kossek and Lee, 2020b; Myers et al., 2020). Studies suggest, too, that team size has decreased during 2020 and that women's shares of first authorships, last authorships, and general representation per author group have decreased during the COVID-19 pandemic (Andersen et al., 2020; Fry et al., 2020).

With variations by discipline, women also published fewer papers and received fewer citations of their work since between March 2020 and December 2020 (Amano-Patino et al., 2020; Andersen et al., 2020; Gabster et al., 2020), which may affect their job stability and future ability to obtain funding. Moreover, the COVID-19 pandemic has exacerbated many stresses women in academia face under usual conditions. For example, delays in obtaining clearance for conducting research during 2020, a result of the COVID-19 pandemic, led researchers to experience increased burnout, sleep disturbance, poor appetite, increased interpersonal problems, and decreased motivation (Sharma et al., 2020).

Available information indicates that the alterations to healthy boundaries between the multiple roles women assume (e.g., as caregivers and professionals) and increased isolation may also negatively impact productivity; harm the recruitment, retention, and persistence of women in STEMM; and/or affect mental well-being. Women faculty in STEMM faced unique challenges resulting from the COVID-19 pandemic related to juggling growing second shift challenges juxtaposed with increased boundary permeability, rising workloads, and persistent ideal worker cultures. While remote work can facilitate the management of work-family roles, it also increases multitasking, process losses from switching frequently between tasks, and interruptions and extended work availability that may harm mental health and well-being. In addition, several studies have shown there are health and well-being implications of these unequal childcare responsibilities (Kossek et al., 2014). To cope with additional caregiving demands, women are reducing their work hours (Madgavkar et al., 2020).

Postsecondary institutions and funders also found themselves in uncharted territory as a result of the COVID-19 pandemic, and they have responded in several ways including altering tenure and promotion policies. Tenure clock

---

[2] Social support—particularly that gained from in-person contact—is a protective factor against the adverse effects of stress on health and during many recent societal stressors in the United States, such as natural disasters and terrorist attacks, individuals have been able to gather with family, friends, and colleagues to grieve and heal.

extensions were widely implemented as policies to address the COVID-19 pandemic productivity challenges[3] and may be important for some faculty members. However, these policies were often implemented without addressing disparities in caregiving and job-related workload that women faculty across all ranks and job status faced. Moreover, extending the tenure clock may put off financial incentives, career advancement, and academic freedom. Many funders modified their policies to allow greater flexibility to researchers in 2020, but there was often no additional funding to support staff and graduate students over the longer project period. While 1-year extensions and grant extension flexibility are helpful, overall, the differential effects for women may not be sufficient to address the added caregiver status and home responsibilities that affect work-life integration.

There were some emerging data by the end of 2020 indicating that approaches some academic leaders used to make decisions, govern, and be accountable were more gender inclusive and may help to eradicate growing equity gaps. The predominant approaches included at least three strategies: utilizing the expertise of existing diversity, equity, and inclusion staff to inform decision-making processes; creating new structures to address decision-making needs; and altering existing processes to include more voices in decision-making. Some campuses began to think about the long-term implications of the COVID-19 pandemic and suggested strategies to address this issue, such as revised strategic plans aimed at ameliorating equity gaps. However, budget cuts made by many colleges and universities in response to the economic constraints that arose during 2020 greatly affected contingent and non-tenured faculty members—positions disproportionately occupied by women and People of Color.

Along with these potential negative effects, the COVID-19 pandemic may be catalyzing changes that could portend a better future for women in academic STEMM. Emerging work from several nations suggest that men have started shouldering more caregiving and child-rearing duties, a view corroborated by their partners (Carlson et al., 2020b; Savage, 2020; van Veen and Wijnants, 2020). A study by a consumer marketing group found that 62 percent of men wanted to keep working at home specifically because it increased family time (Fluent, Inc., 2020). As the collective boundaries and barriers between work and home were removed, there were instances of humanization and increased understanding or empathy. Professional conferences adapted quickly to virtual platforms, allowing global participation and often increasing access by removing travel-related barriers that can affect women more than men, given their caregiving responsibilities. Taken together—positive and negative—it is important to identify and illuminate the ways that the COVID-19 pandemic has affected and will affect women in academic STEMM for years to come.

---

[3] An informally gathered list of changes to tenure clock policies is available at https://docs.google.com/spreadsheets/d/1U5REApf-t-76UXh8TKAGoLlwy8WIMfSSyqCJbb5u9lA/edit#gid=0&fvid=238051147.

THE IMPORTANCE OF INTERSECTIONALITY AND EQUITY

The concept of intersectionality—a lens for understanding how social identities, especially for marginalized groups, relate to systems of authority and power—is helpful for understanding how the COVID-19 pandemic continues to affect women in STEMM. Productivity, careers, boundary setting, and mental health and well-being are all influenced by the ways in which social power structures define and cultivate social identities. Race and ethnicity, sexual orientation, gender identity, age, and disability status, among many other factors, can amplify or alter the effects of the COVID-19 pandemic for a given person. It is therefore critical to investigate, understand, and present the topics explored herein through an equity lens. However, while international scientists have been specifically affected by travel restrictions, increased isolation, and other impacts of the COVID-19 pandemic, this report does not focus on the citizenship of the researchers when considering intersectionality and equity (OECD, 2020).

OTHER SIGNIFICANT FACTORS

Although the primary focus of this study is the COVID-19 pandemic, it was not the only crisis that affected the United States in 2020 and could not be considered in isolation. The committee considered several contextual elements that interacted extensively with the COVID-19 pandemic, including the effects of anti-Black racism, the economic recession triggered by the COVID-19 pandemic, and the increase in technology-mediated interactions, all of which had additional significance for the careers of women in academic STEMM.

Difficult and compounding issues, such as the persistence of structural injustices in U.S. society and increasing frequency and severity of natural disasters, led to the creation of departmental committees, special councils, and task forces. Understanding the implications of these issues will take time and careful examination. A complete review lies beyond the scope of this report, but it is important to recognize that these events and activities were occurring and absorbing resources of administrators and faculty—particularly Faculty of Color—concurrently with the COVID-19 pandemic. The devastating economic recession not only affected women in STEMM at postsecondary institutions directly, as they felt the consequences of policies such as furloughs, hiring freezes, and elimination of merit increases, but also indirectly as women were forced out of the workforce at a higher rate than men and were more likely to report the need to leave the workforce if their children's school systems did not have in-person classes in the fall of 2020 (FRB, 2020; Heggeness, 2020). The increased use of technology presented many opportunities for increased access and engagement as well as potential avenues for virtual harassment. It also required resources (financial and technological) to engage with the various platforms that are not ubiquitous in the United States.

BEYOND 2020

Expectations of gendered roles in society often take on added emphasis during periods of stress, and there is a risk that the divide between those who have privilege and those who do not may deepen. Explicit attention to the early indicators of how the COVID-19 pandemic is affecting women in academic STEMM careers, as well as attention to crisis responses throughout history, may illuminate opportunities to mitigate some of the long-term effects and even create a more equitable system.

The information and experiences assembled in this report represent a description of what was known by the end of 2020. While the report may help inform the decisions that academic leaders, funders, other interested stakeholders, and both current and aspiring academics will continue to have to make over the course of the COVID-19 pandemic, the charge to the committee was to inform, without making recommendations. Academic leaders and key decision makers may use the information gathered in this study as they consider new policies or adapt current ones to be more responsive to the challenges that women in academic STEMM experience. They may also use the report's findings as they look for new ways to engage and move forward in creating a more equitable and inclusive higher education and research system. In addition, the lessons that can be gleaned from the first several months of the COVID-19 pandemic may be applicable to other large-scale disruptions (e.g., climate change–related events, severe economic recessions, or other novel infectious disease outbreaks) that will continue to be risks faced by the STEMM enterprise over time.

This report, however, was motivated by and focused on the COVID-19 pandemic. Building on the *Promising Practices* report, the committee provides findings based on the preliminary evidence available during 2020 and identifies questions to create a research agenda about short- and long-term impacts of the COVID-19 pandemic. Together, the findings and research questions can help better prepare higher education institutions to respond to disruptions and explore opportunities that support the full participation of women in the future.

The future almost certainly holds additional, unforeseen disruptions that will test the principles and resilience of institutions of higher education. It also almost certainly requires the contributions of STEMM, which can be fully realized only if the well-being of women in these fields does not significantly suffer from the COVID-19 pandemic and other disruptions.

MAJOR FINDINGS

Given the ongoing nature of the COVID-19 pandemic, it was not possible to fully understand the entirety of the short- or long-term implications of this global disruption on the careers of women in academic STEMM. Having gathered preliminary data and evidence available in 2020, the committee found that significant changes to women's work-life boundaries and divisions of labor, careers, productivity, advancement, mentoring and networking relationships, and mental health and well-being have been observed. The following findings represent those aspects that the committee agreed have been substantiated by the preliminary data, evidence, and information gathered by the end of 2020. They are presented either as Established Research and Experiences from Previous Events or Impacts of the COVID-19 Pandemic during 2020 that parallel the topics as presented in the report.

Established Research and Experiences from Previous Events

Finding 1 **Women's Representation in STEMM:** Leading up to the COVID-19 pandemic, the representation of women has slowly increased in STEMM fields, from acquiring Ph.D.s to holding leadership positions, but with caveats to these limited steps of progress; for example, women representation in leadership positions tends to be at institutions with less prestige and fewer resources. While promising and encouraging, such progress is fragile and prone to setbacks especially in times of crisis (see Chapter 6).

Finding 2 **Confluence of Social Stressors:** Social crises (e.g., terrorist attacks, natural disasters, racialized violence, and infectious diseases) and COVID-19 pandemic-related disruptions to workload and schedules, added to formerly routine job functions and health risks, have the potential to exacerbate mental health conditions such as insomnia, depression, anxiety, and posttraumatic stress. All of these conditions occur more frequently among women than men.[4] As multiple crises coincided during 2020, there is a greater chance that women will be affected mentally and physically (see Chapters 4 and 7).

Finding 3 **Intersectionality and Equity:** Structural racism is an omnipresent stressor for Women of Color, who already feel particularly isolated in many fields and disciplines. Attempts to ensure equity for all women may not necessarily create equity for women across various identities if targeted interventions designed to promote gender equity do not account for the racial and ethnic heterogeneity of women in STEMM (see Chapters 1, 3, and 4).

---

[4] This finding is primarily based on research on cisgender women and men.

Impacts of the COVID-19 Pandemic during 2020

Finding 4 **Academic Productivity:** While some research indicates consistency in publications authored by women in specific STEMM disciplines, like Earth and space sciences, during 2020, several other preliminary measures of productivity suggest that COVID-19 disruptions have disproportionately affected women compared with men. Reduced productivity may be compounded by differences in the ways research is conducted, such as whether field research or face-to-face engagement with human subjects is required (see Chapter 3).

Finding 5 **Institutional Responses:** Many administrative decisions regarding institutional supports made during 2020, such as work-from-home provisions and extensions on evaluations or deliverables, are likely to exacerbate underlying gender-based inequalities in academic advancement rather than being gender neutral as assumed. For example, while colleges and universities have offered extensions for those on the tenure track and federal and private funders have offered extensions on funding and grants, these changes do not necessarily align with the needs expressed by women, such as the need for flexibility to contend with limited availability of caregiving and requests for a reduced workload, nor do they generally benefit women faculty who are not on the tenure track. Furthermore, provision of institutional support may be insufficient if it does not account for the challenges faced by those with multiple marginalized identities (see Chapters 3 and 4).

Finding 6 **Institutional Responses:** Organizational-level approaches may be needed to address challenges that have emerged as a result of the COVID-19 pandemic in 2020, as well as those challenges that may have existed before the pandemic but are now more visible and amplified. Reliance on individual coping strategies may be insufficient (see Chapters 2 and 6).

Finding 7 **Work-Life Boundaries and Gendered Divisions of Labor:** The COVID-19 pandemic has intensified complications related to work-life boundaries that largely affect women. Preliminary evidence from 2020 suggests women in academic STEMM are experiencing increased workload, decreased productivity, changes in interactions, and difficulties from remote work caused by the COVID-19 pandemic and associated disruptions. Combined with the gendered division of nonemployment labor that affected women before the pandemic, these challenges have been amplified, as demonstrated by a lack of access to childcare, children's heightened behavioral and academic needs, increased eldercare demands, and personal physical and mental health concerns. These are particularly salient for women who are parents or caregivers (see Chapter 4).

Finding 8 **Collaborations:** During the COVID-19 pandemic, technology has allowed for the continuation of information exchange and many collaborations. In some cases technology has facilitated the increased participation of women and underrepresented groups. However, preliminary indicators also show gendered impacts on science and scientific collaborations during 2020. These arise because some collaborations cannot be facilitated online and some collaborations face challenges including finding time in the day to engage synchronously, which presents a larger burden for women who manage the larger share of caregiving and other household duties, especially during the first several months of the COVID-19 pandemic (see Chapter 5).

Finding 9 **Networking and Professional Societies:** During the COVID-19 pandemic in 2020, some professional societies adapted to the needs of members as well as to broader interests of individuals engaged in the disciplines they serve. Transitioning conferences to virtual platforms has produced both positive outcomes, such as lower attendance costs and more open access to content, and negative outcomes, including over-flexibility (e.g., scheduling meetings at non-traditional work hours; last-minute changes) and opportunities for bias in virtual environments (see Chapter 5).

Finding 10 **Academic Leadership and Decision-Making:** During the COVID-19 pandemic in 2020, many of the decision-making processes, including financial decisions like lay-offs and furloughs, that were quickly implemented contributed to unilateral decisions that frequently deviated from effective practices in academic governance, such as those in crisis and equity-minded leadership. Fast decisions greatly affected contingent and nontenured faculty members—positions that are more often occupied by women and People of Color. In 2020, these financial decisions already had negative, short-term effects and may portend long-term consequences (see Chapter 6).

Finding 11 **Mental Health and Well-being:** Social support, which is particularly important during stressful situations, is jeopardized by the physical isolation and restricted social interactions that have been imposed during the COVID-19 pandemic. For women who are already isolated within their specific fields or disciplines, additional social isolation may be an important contributor to added stress (see Chapter 7).

Finding 12 **Mental Health and Well-being:** For women in the health professions, major risk factors during the COVID-19 pandemic in 2020 included unpredictability in clinical work, evolving clinical and leadership roles, the psychological demands of unremitting and stressful work, and heightened health risks to family and self (see Chapter 7).

RESEARCH QUESTIONS

While this report compiled much of the research, data, and evidence available in 2020 on the effects of the COVID-19 pandemic, future research is still needed to understand all the potential effects, especially long-term implications. The research questions represent areas the committee identified for future research, rather than specific recommendations. They are presented in six categories that parallel that chapters of the report: Cross-Cutting Themes; Academic Productivity and Institutional Responses; Work-Life Boundaries and Gendered Divisions of Labor; Collaboration, Networking, and Professional Societies; Academic Leadership and Decision-Making; and Mental Health and Well-being. The committee hopes the report will be used as a basis for continued understanding of the impact of the COVID-19 pandemic in its entirety and as a reference for mitigating impacts of future disruptions that affect women in academic STEMM. The committee also hopes that these research questions may enable academic STEMM to emerge from the pandemic era a stronger, more equitable place for women. Therefore, the committee identifies research questions for both understanding the impacts of the disruptions from the COVID-19 pandemic and exploring the opportunities to help support the full participation of women in the future.

Research questions listed under Cross-Cutting Themes tackle the overall impacts of the COVID-19 pandemic on women's participation in STEMM, particularly Women of Color, and the confluence of social stressors that were experienced during 2020. Academic Productivity and Institutional Response research questions probe how policies and practices such as extensions may affect the career trajectories of women, both short and long term. The challenges and insights gained from work-life management, especially for parents and other caregivers, are explored in the Work-Life Boundaries and Gendered Division of Labor research questions. Collaboration, Networking, and Professional Societies research questions cover several aspects of the changed nature of connectivity and conferencing. Research questions about models of leadership and potential institutional change are included under Academic Leadership and Decision-Making. Finally, research questions listed under Mental Health and Well-being consider how colleges and universities can support their academic STEMM workforce. during and after societal stressor events like the COVID-19 pandemic. The full list of research questions is provided in Chapter 8.

1

Introduction

Emerging evidence suggests that the COVID-19 pandemic endangers the engagement, experience, and retention of women in science, technology, engineering, mathematics, and medicine (STEMM). If these possibilities are not outlined systematically and addressed intentionally, the COVID-19 pandemic may have long-term impacts on the future of the STEMM workforce. Moreover, if new opportunities for improving the engagement and retention of women that have arisen during the course of 2020 are not explored, the capacity for building, learning, and developing improved systems may diminish. Thus, just as the 2020 National Academies of Sciences, Engineering, and Medicine consensus study report *Promising Practices for Addressing the Underrepresentation of Women in Science, Engineering, and Medicine: Opening Doors* (the *Promising Practices* report) called attention to the challenges that women in STEMM experience and presented evidence-based recommendations to address the well-established structural barriers that impede the advancement of women in STEMM (NASEM, 2020), this committee identified and illuminated the ways that the COVID-19 pandemic affected women in academic STEMM during 2020, and laid out a path for future research for years to come.

The reason it is important to understand the potential effects of the COVID-19 pandemic on women is that there is clear evidence that advances in knowledge and practice in the fields of academic STEMM benefit from a workforce composed of people who represent a diversity from multiple perspectives, including gender, ethnicity, and ancestry. These perspectives contribute unique "vectors of skills, experiences, and talents" to the STEMM enterprise (Page, 2007, 2008, 2019). Despite these observed benefits, there remains a paucity of women in

STEMM, with both structural and institutional inequities contributing to women's persistent underrepresentation and undue burdens in academic STEMM fields.[1] Recognizing the importance of diversity to knowledge production, public and private agencies have invested heavily in efforts to diversify STEMM fields and, over the past few decades there has been an absolute increase in the number of women—particularly white women and, to a degree, Asian women—who have earned degrees across the STEMM landscape. Although progress has been slower for Black, Indigenous, and Latina women, there is no denying that the presence of Women of Color in STEMM is expanding. However, given that STEMM fields generally grant privilege to white, masculine norms, these advances in diversity are not secure.

REPORT CONTEXT

To fully understand the context of this report, it is important to recall how the COVID-19 pandemic upended almost all workplaces during 2020, including higher education, where much STEMM work takes place. In March 2020, when public health experts and U.S. government officials recognized the gravity of the global pandemic, almost every college and university restricted in-person activities such as teaching, research, and laboratory operations (Holshue et al., 2020).[2] Faculty members, many of whom had never experienced online education as instructors nor as students (Terosky and Gonzales, 2016), shifted to online instruction; suspended human, laboratory, and field research; and discontinued or adapted their service and mentorship work to a new mode of operation. Similarly, academic medical centers shifted focus to urgent operations in the face of tremendous clinical demands while sustaining medical education through virtual curricula. Almost immediately, the COVID-19 pandemic began to affect the finances of colleges, universities, and academic medical centers; student access and experiences; and faculty careers. Early in the COVID-19 pandemic, it seemed likely that there were disparate and potentially lasting effects on women faculty, staff, and scholars in higher education, especially women in STEMM (Collins et al., 2020; Schreiber, 2020). For example, many women reported a rapid expansion of caregiving and domestic tasks, relative to pre-COVID-19 pandemic levels, coupled with the quickly evolving and emotional demands related to virtual forms

---

[1] Throughout this report, the term *women* is understood to broadly include all individuals who identify as women. Furthermore, the committee is focused on gender, rather than sex, prompting the use of "woman," not "female." Woman refers to a person's gender, whereas female refers to a person's sex. Although these terms are often used interchangeably and can be related, they are discrete concepts. Gender is a nonbinary social construct, whereas sex is primarily considered a biological trait.

[2] The first recognized detection of SARS-CoV-2, the virus that led to the COVID-19 pandemic, in the United States was recorded in January 2020. The first case in China occurred in December 2019 (Holshue et al., 2020). Although some academic institutions did not give faculty members any choice in teaching mode, many provided options for faculty who are in higher-risk brackets for COVID-19 to teach remotely or in person as well as resources for faculty who need to teach from home.

of teaching, mentoring, and research (Cardel et al., 2020a; Gabster et al., 2020; Power, 2020). Throughout the course of 2020, these expectations did not subside: women scholars continued to report tensions related to caregiving, household management, and work, and caregiving was no longer part of a "second shift" job, but concomitant with the "first shift" of their academic job.

The effects of COVID-19 during 2020 varied across STEMM disciplines. Laboratory-based researchers reported being unable to do work due to the shutdown of laboratories (Korbel and Stegle, 2020), whereas researchers in computational biology, data science, mathematics, statistics, computer science, and economics reported being able to continue work and collaborations, and reported minimal decline in research time (Korbel and Stegle, 2020; Myers et al., 2020). Some fields, like the geosciences, reported increases in paper submissions in 2020 (Wooden and Hanson, 2020). One survey indicated that many life scientists took the opportunity to analyze data, write manuscripts or theses, or improve skills through online coursework (Korbel and Stegle, 2020).

As summarized in the *Promising Practices* report, it is challenging to address and dismantle the persistent inequities experienced by women in academia (Bickel, 2004; Holman et al., 2018; Mangurian et al., 2018; NASEM, 2020). As that report concluded, bias, discrimination, and harassment are major contributors to the underrepresentation of women; it recommended that institutions create a supportive working environment by allocating resources to support research, teaching, advancement, and career development, and to create institutional structures that promote fairness and transparency. Overcoming unconscious bias is linked to slowed decision-making (NASEM, 2020).[3] However, given the accelerated time frames that many leaders faced during the initial response to the COVID-19 pandemic in 2020 to transform their educational, research, and clinical environments, some of these deliberately slow processes might have seemed impossible to implement.

Recognizing the strain of the COVID-19 pandemic on faculty, many colleges and universities quickly adopted policies such as tenure clock extensions to relieve the pressure on pre-tenure colleagues. However, research shows that while tenure clock extensions may be helpful to some, it is not a policy solution that attends to all (Antecol et al., 2018; Gonzales and Griffin, 2020; Scott, 2016). Importantly, the vast majority of faculty do not hold tenure-track appointments (Curtis, 2019). Instead, non-tenure-track or contingent faculty, clinical faculty, lecturers, research associates and scientists, and postdoctoral scholars together account for about 73 percent of the entire academic workforce, including in STEMM fields (AAUP, 2018, 2020a, 2020b; Curtis, 2019; Finkelstein et al., 2016; Kezar et al., 2019). Moreover, women, and particularly Black, Indigenous,

---

[3] For example, when hiring, best practices include posting all positions; using carefully considered language in the job description; reflecting on and ranking the specific qualifications that are expected and criteria that will be evaluated; encouraging search committees to recruit broadly and creatively; and training members to mitigate their own biases (NASEM, 2020).

and Latina women, are more likely to hold non-tenure-track faculty positions, implying that tenure clock extensions have no bearing on their work arrangements and that they are also more likely to suffer temporary or permanent layoffs (Chronicle Staff, 2020; Kalev, 2020; Kezar et al., 2019; Nzinga, 2020; Pettit, 2020b).[4] Even for those on the tenure track, there is evidence that in some fields, gender-neutral tenure-track extension policies actually widened the gap between the tenure rates of women and men by increasing the tenure rate for men while decreasing it for women (Antecol et al., 2018). For women striving to establish themselves as STEMM professionals, hiring freezes at universities triggered by the economic fallout resulting from the COVID-19 pandemic may also disrupt the short-term career flow of Ph.D. candidates and postdoctoral scholars, as well as the career aspirations of some tenure-track and tenured faculty and contract renewal prospects for non-tenure-track faculty and staff.

Research suggests that when women are able to return to their professional responsibilities, they may face heightened demands for support from students as well as colleagues (Gonzales and Griffin, 2020). Before the COVID-19 pandemic, scholars established that women faculty do more work involving significant emotional labor, which includes affirming and mentoring students as well as early-career or peer colleagues, and identifying key resources for these individuals (Armstrong and Jovanovic, 2015, 2017; Bellas, 1999; Ruder et al., 2018; Smith, 2019; Turner and González, 2011).

For academic Women of Color, especially Black women, the compounding effects of racism during the COVID-19 pandemic in 2020 cannot be ignored. Many Women of Color often face extraordinary expectations as one of a few Faculty of Color, such as the perception that they can and do represent and serve all People of Color. For Women of Color, research suggests explicitly that the intersection of their racial and gender identities multiplies the expectations for emotional labor and care work, as Students of Color and women students seek out guidance from relatable role models (Moore et al., 2010; Porter et al., 2018). Although this work is incredibly important, it largely goes unrecognized in policies and processes related to career advancement (Bellas, 1999; Gonzales and Ayers, 2018; Hanasono et al., 2019; Oleschuk, 2020; Power, 2020; Tunguz, 2016). If institutions fail to recognize these additional dimensions of labor, women, and particularly Women of Color, may become emotionally depleted, burned out, and in need of additional support for their well-being, with less time and energy to commit to it. During the COVID-19 pandemic, more than a quarter of women in the workforce considered reducing their workloads, shifting their careers, or leaving the workforce (Coury et al., 2020). Although this finding was based on women working within the private corporate sector, there are similar stories in higher education news and scholarly outlets (Cardel et al., 2020a; McMurtrie,

---

[4] An informally gathered list of changes to tenure clock policies is available at https://docs.google.com/spreadsheets/d/1U5REApf-t-76UXh8TKAGoLlwy8WIMfSSyqCJbb5u9lA/edit#gid=0&fvid=238051147.

2020; Somerville and Gruber, 2020; Woolston, 2020a). This report arises out of the need to expeditiously identify, name, and document how the COVID-19 pandemic affected the careers of women in academic STEMM during the initial 9-month period since March 2020 and what the potential downstream effects might be. Preliminary evidence indicates that disruptions in 2020 could have both short- and long-term consequences that will likely vary across institution type (e.g., community colleges, baccalaureate-granting institutions, doctoral-granting and research universities, Historically Black Colleges and Universities, Hispanic-Serving Institutions, and Tribal Colleges and Universities); career stage or focus (e.g., graduate student; postdoctoral scholar; medical resident; clinician; tenure-track, tenured, full-time non-tenure-track, and adjunct faculty); academic rank (e.g., assistant professor, associate professor, full professor); and personal characteristics, including family structure, caregiving responsibilities, and behavioral health status.

BEYOND THE COVID-19 PANDEMIC

The COVID-19 pandemic was devastating in so many ways, but it was not the only major event affecting the nation in 2020. Although the primary focus of this study is the COVID-19 pandemic's potential impacts on women's STEMM career trajectories, the committee recognized several contextual elements that actively intensified the immediate effect of the COVID-19 pandemic, including the effects of anti-Black racism, the economic recession triggered by the COVID-19 pandemic, and the sudden importance of technology-mediated interactions. While this report was motivated by and focuses almost entirely on the COVID-19 pandemic, other large-scale disruptions (e.g., climate change–related events, severe economic recessions, or other novel infectious disease outbreaks) will continue to be risks that the scientific enterprise faces. In that regard, this report may prove useful in responding to future crises.

Racism

While racism is intertwined in every aspect of American life and the history of racism in the United States is well documented, the summer of 2020 marked an awakening for many people, particularly white people, about the deep historical roots and pervasive nature of racism in this country. It is important to recognize and understand the always-present impacts of discriminatory practices and behavior—particularly anti-Black and anti-Indigenous racism—in the United States and globally. Indeed, even when looking at the demographics of lives lost to the COVID-19 pandemic, there are wide inequities observed by race and socioeconomic factors, especially for Black, Latinx, and Indigenous Americans (APM Research Lab Staff, 2020). For example, the Centers for Disease Control and Prevention's November 30, 2020, update on ratios of hospitalization and death by race/ethnicity demonstrated that Hispanic/Latinx persons, non-Hispanic American Indian/Alaska Native persons, and non-Hispanic Black/African American persons experienced between 1.4 and 1.8 times as many cases, approximately 4 times the number of hospitalizations, and more than 2.5 times as many deaths compared with white, non-Hispanic individuals (CDC, 2020a).[5]

Black women incurred unique burdens during 2020. Articles from June and October 2020 described the increased stress and terror that some Black women experienced as a result of being immersed in a climate with persistent racism, being the target of anti-Black sentiments, and feeling compelled to speak out to validate and support Black Lives Matter activities, all compounded by effects of the COVID-19 pandemic (Flaherty, 2020g; McCoy, 2020). Because the COVID-19 pandemic has led to disproportionate numbers of Black people becoming ill or dying, Black women also report feelings of deep despair as the virus impacts the lives of friends and relatives or their own health (McCoy, 2020).

One outcome of the increased attention and public discussion of systemic racism in the United States during the summer of 2020 was an increased focus on anti-racism as an approach to addressing, preventing, or mitigating harms within university systems (Belay, 2020; Gray et al., 2020; Taylor Jr. et al., 2020). Contrary to being not racist, which is a neutral stance that can nonetheless perpetuate racism through inaction, a person or institution that is anti-racist actively addresses and supports policies that dismantle inequities between all marginalized populations (Kendi, 2019). Taking an anti-racist approach to understanding university systems and supports requires institutions to scrutinize the systemic racial bias that is endemic within the fabric of their own organizational culture, specifically policies and practices and the communities served, as well as nationally.

Often, workplace diversity practices do not analyze racial oppression or commit to disrupting it from a structural perspective (Ahmed, 2012; Zschirnt, 2016).[6] Research does indicate that proactively hiring diverse staff and leadership across many attributes—ancestry, gender, sexual orientation, and personality type—results in increased levels of creativity, innovation, and effective problem solving (Page, 2007, 2008, 2019). By understanding racial oppression as a structural occurrence, rather than the result of individual acts of prejudice, it is possible to confront the root causes of racial disparities (Cassedy, 1997; McSwain, 2019; Pager and Shepherd, 2008). Therefore, the committee could not ignore how racism and gender come together to uniquely affect Black women, among other Women of Color, in the context of this study. In that regard, the committee used the concept of intersectionality—a lens for understanding how social identities, especially for marginalized groups, relate to systems of authority and power—to help it examine the possible ways that the COVID-19 pandemic is affecting Women of Color specifically.

---

[5] Ratios are given as age-adjusted rates standardized to the 2000 U.S. standard population.

[6] *Workplace diversity* refers to demographic variation within an organization's staff and leadership.

Economy

Gender-specific disruptions resulting from the COVID-19 pandemic are apparent in virtually every sector of the economy (Alon et al., 2020a, 2020b). Overall, women have been forced out of the workforce at a higher rate than men and were more likely to report they will need to leave the workforce if their children's school systems did not have in-person classes in fall 2020 (FRB, 2020; Heggeness, 2020). Disconnecting women in academia from these household issues is simply not possible. As is discussed throughout this report, these economy-wide effects on women are likely a result of the historic gender imbalance of household and caregiving duties persisting into the COVID-19 pandemic (Miller, 2020).

It was clear before the pandemic that one of the most important challenges to a convergence in opportunities for persons of any gender in the workplace was a need for flexibility—in compensation schemes, work arrangements, and scheduling, among other considerations (Goldin, 2014). On one hand, the inflexibility of many COVID-19 pandemic-related policies (e.g., social distancing), alongside women's greater willingness to abide by these policies, is proving a challenge for women in virtually every industry (Coury et al., 2020). On the other hand, the growth of more flexible work arrangements in remote work environments and with dynamic work schedules could prove to be important testbeds for the technologies and work-related contracts that have long been needed to allow for equitable opportunities for all genders in all workplaces.

Technology

Technology has assumed mixed roles during the COVID-19 pandemic, especially as students, faculty, researchers, and others attempt to restore work activity. In some respects, technology has facilitated increased access to conferences, collaborations with scholars outside their institutions, and delivery of content for courses in novel and creative ways (Gottlieb et al., 2020; Niner et al., 2020). However, technology has also resulted in creating workplaces that often have less clarity related to boundaries between personal and professional lives. It has forced households that typically had one or two individuals using the internet at the same time to adjust to a new reality in which adults engaged in work and students involved in classwork are competing for bandwidth, forcing families to make larger investments in internet capacity (Bacher-Hicks et al., 2020; Stelitano et al., 2020; Vogels et al., 2020). For many, especially contract-based faculty, undergraduate and graduate students, and postdoctoral scholars, economic barriers can strain equitable access to sufficient Wi-Fi or other internet access. Thus, whereas this may seem like a minor inconvenience for some academics, it can be consequential for others (Zahneis, 2020).

Finally, women in STEMM face excessive sexual harassment and other forms of discrimination (NASEM, 2018), a phenomenon not isolated to in-person interactions (Barak, 2005). A key contribution of the 2018 National Academies report on harassment of women was to point out that sexual harassment most commonly involves sexist remarks and behaviors, something that in-person interactions are not necessary to perpetuate. Technology can provide a harasser with additional access to a target, including the ability to monitor a target's whereabouts and be in constant communication, which may exacerbate feelings of powerlessness for a target (Southworth et al., 2007). Some evidence suggests that the impact of sexual harassment in the online setting (Gáti et al., 2002) mirrors that experienced offline (Harned and Fitzgerald, 2002).

STUDY PROCESS

In August 2020, the National Academies assembled an ad hoc study committee to build on the *Promising Practices* report and examine early indicators of the impacts of the COVID-19 pandemic on the careers of women in academic STEMM fields. The full Statement of Task for the committee is provided in Box 1-1.

Recognizing the preliminary nature of the data and evidence available during the course of the study, the committee discussed several key aspects regarding the interpretation of the Statement of Task. The committee also recognized the importance of defining key terms and their consistent use in the study process and report (see Box 1-2 and Glossary). In particular, the committee interpreted "STEMM academic careers" to include all faculty (tenure-track and non-tenure-track), postdoctoral scholars, and graduate students, and STEMM was determined to reference

---

BOX 1-1
Statement of Task

An ad hoc committee of the National Academies of Sciences, Engineering, and Medicine will undertake a fast-track study focused on early indicators of the potential impact of the COVID-19 pandemic on the careers of women in academic science, engineering, and medicine (STEMM). Building on the information and data in the recent National Academies report *Promising Practices for Addressing the Underrepresentation of Women in STEMM*, the committee will commission research papers to identify and analyze disruptions experienced by women in STEMM academic careers during the early stages of the COVID-19 pandemic. It will also hold a public workshop and conduct its own analyses on the ways in which COVID-19 is amplifying the disruptions encountered by women in STEMM academic careers, such as those related to child and family caregiving responsibilities. Based on the commissioned papers, workshop presentations, and its own analyses, the committee will issue a consensus report with findings that reflect what has been learned through its work as well as recommendations for further study and investigation.

> **BOX 1-2**
> **On the Use of Language**
>
> During the study process and drafting of the report, the committee recognized the importance of defining terms that may be unfamiliar to some readers or are used variably across contexts (see Glossary for full list). Four terms used in this report merit particular attention.
>
> - **Academic STEMM workforce**, for purposes of this report, includes all individuals in a STEMM field employed at a college or university in an academic position, including tenure-track and non-tenure-track faculty, among others with teaching, research, clinical, outreach, extension, or other "engagement" responsibilities. Where noted, this may also include postdoctoral researchers and graduate students.
> - **Intersectionality** is a field of study and an analytic lens that makes visible the mutually constructive and reciprocal relationship among race, ethnicity, sexuality, class, and other social positions that influence one's experiences (Collins, 2015). In this report, it is used as a lens to help understand how social identities, especially for marginalized groups, relate to systems of authority and power. Intersectionality is rooted in Black feminism and Critical Race Theory: in reference to historic exclusion of Black women, legal scholar Kimberlé Williams Crenshaw used intersectionality to describe the intersection of gender and race discrimination, arguing that treating them as exclusive, and not intertwined, renders the multiple marginalities faced by Black women invisible to antidiscrimination law (Carbado, 2013; Crenshaw, 1989, 1991, 2014).
> - **People of Color** is an evolving term that emerged in the 1960s, which now includes a broader group of individuals such as Black, Latinx, Asian, Mexican, Japanese, Chinese, and other groups that share a common societal place of either feeling or actually being marginalized (Pérez, 2020). For purposes of this report, People of Color refers to all individuals specifically identifying as Black, Latinx, and American Indians/Alaska Natives.
> - **Woman** refers to any person who identifies as a woman, including, but not limited to, cisgender women, transgender women, and nonbinary women. "Woman" refers to a person's gender, whereas "female" refers to a person's sex. Although these terms are often used interchangeably and can be related, they are discrete concepts. The committee uses gender-first language except where specifically noted that the research discussed is referring to sex.

all sciences, engineering, mathematical, and medical fields.[7] The committee strove to solicit balanced information regarding both positive and negative impacts of the disruptions of the COVID-19 pandemic, seeking not just to illuminate hardships but also to highlight potential opportunities for growth or innovation. The preponderance of the preliminary data and evidence available at the end of 2020, however, indicate overall negative effects on the careers of women in academic STEMM during the first several months of the COVID-19 pandemic.

In addition, much of the preliminary evidence available on the effects of the COVID-19 pandemic on women in academic STEMM may be evidence of the effects on all women, particularly working women. In many cases, the effects described also affected men in academic STEMM. Where possible, care has been taken to acknowledge if the data and evidence are only available for women in academic STEMM, or if the evidence provided supports findings for a broader population. If available, comparative resources that may provide insights into differential impacts by gender, context, or other defining attribute are discussed. However, because it fell outside of the Statement of Task, the committee did not specifically pursue research on the differential impacts on academic men or on nonacademic women.

The *Promising Practices* report provided nuanced observations about both common traits and unique characteristics of different fields within STEMM, as well as detailed overarching recommendations. This committee strove to provide similar insights from the data and evidence available in 2020. While some distinctions are described throughout the report, the committee acknowledges that more information and understanding about how women in different fields within STEMM were affected during 2020 and beyond will come with time.

The *Promising Practices* report also emphasized the importance of understanding the intersectional identities of Women of Color and women from other marginalized groups.[8] The committee therefore also recognized the importance of understanding the effects of the COVID-19 pandemic on women in academic STEMM through an intersectional lens. As discussed previously, this was additionally important because of the concurrent events that elevated the discussions around racism, especially anti-Black racism, during 2020.

Commissioned Papers

To inform its deliberations, findings, and research questions, the committee commissioned five papers. Each paper focused on a unique aspect of how the COVID-19 pandemic has affected women STEMM academics during 2020. At

---

[7] The committee determined that the specific nuances of careers in technology were being addressed by other studies. The committee, however, decided to use the acronym "STEMM" throughout the report.

[8] Intersectionality is a lens or analytical framework for understanding how aspects of an individual's social identities, especially for marginalized groups, relate to systems of authority and power and the complexity of the prejudices they face.

the time the papers were commissioned, the committee reviewed preliminary evidence on the potential impact of the COVID-19 pandemic. Furthermore, the committee wanted to explore responses to the pandemic that did not exacerbate preexisting issues (many of which are highlighted in the *Promising Practices* report), or lead to regression of limited gains in diversity, equity, and inclusion made over the past several decades, and that could identify opportunities and potentially lead to more resilient and equitable higher education systems. The five overarching topics of the papers as commissioned (and their authors) are as follows: the Impact of COVID-19 on (1) Tenure Clocks, the Evaluation of Productivity, and Academic STEMM Career Trajectories (Felicia A. Jefferson, Matthew T. Hora, Sabrina L. Pickens, and Hal Salzman); (2) Boundary Management, Work-Life Integrations, and Domestic Labor (Ellen Kossek, Tammy D. Allen, and Tracy Dumas); (3) Collaboration, Mentorship and Sponsorship, and the Role of Networks and Professional Organizations (Rochelle Williams and Misty Heggeness); (4) Academic Leadership and Decision-Making (Adrianna Kezar); and (5) the Mental Health and Well-being of Women in STEMM (C. Neill Epperson, Elizabeth Harry, Judith G. Regensteiner and Angie Ribera).

The following six chapters are based on the final drafts of these five papers. Each chapter provides key insights about how the COVID-19 pandemic has affected the careers of women in academic STEMM fields. Chapter 2 sets the stage for the ensuing chapters and presents the results of a survey conducted in October 2020, providing a window into the very personal perspectives offered by respondents;[9] Chapters 3 through 7 approach the core concept of how the COVID-19 pandemic has affected the careers of women in academic STEMM fields from different disciplinary perspectives. Chapters 3 through 7 each review literature and concepts established before the COVID-19 pandemic, summarize the preliminary evidence and data on the impacts of the COVID-19 pandemic during 2020 from the perspective of that field, and—where possible—speculate about potential long-term implications. Taken together, these six different approaches form a single, unified description of the potential impacts of the COVID-19 pandemic on the careers of women in STEMM during 2020.

Cross-Cutting Themes and the Use of Data and Evidence

Beyond their core focus, each commissioned paper also considered a set of cross-cutting themes that the committee identified: historical perspectives that may inform the current and future contexts; national focus on racism; the role of technology throughout the COVID-19 pandemic; and the organizational and operational contexts of colleges and universities, including financial, leadership, and accountability. The commissioned authors were also asked to take a distinctively intersectional lens, although the preliminary data and evidence were not

---

[9] Details about the survey are provided in Chapter 2 and Appendix A. The survey was designed and fielded by Ellen Kossek, Tammy D. Allen, and Tracy Dumas.

always available for this approach. Although each paper may not have been able to address all the themes, the committee considered them collectively to best understand the impact of the COVID-19 pandemic on women in academic STEMM.

All the commissioned paper authors were asked to provide explicit descriptions of the sources of data and evidence that they used in writing their papers, including the considerations they used to select or exclude sources and how they analyzed the evidence. This request was made because of the overall recognition by the committee that the data and evidence available to the authors are still emerging. The committee did not, however, specify or limit the authors as to how they chose their data sources and welcomed well-documented quantitative and qualitative scholarship, as well as historically relevant information.

Each paper was presented for public discussion during a series of webinars in November 2020.[10] In the final chapter of this report, the committee has elevated major findings from across all the chapters, as well as research questions that the committee determined to be of key importance to furthering the overall understanding of the effects of the COVID-19 pandemic on the academic research careers of women in STEMM and the potential opportunities to create a more equitable and resilient higher education and research system. Given the preliminary and evolving nature of the data and evidence available during 2020, the committee was not tasked with and does not make recommendations for policy changes or actions individuals or institutions should take.

ABOUT THE REPORT

This report is intended to provide nuanced insights and "grounding" about the COVID-19 pandemic's varied effects on women in STEMM across different institutional types and at different stages of their career as understood at the end of 2020. Academic leaders and key decision makers may be able to use the preliminary information gathered through this study and laid out in this report to inform new policies or adapt current ones to be more responsive to the challenges that women in academic STEMM are experiencing. The preliminary information and experiences assembled in the subsequent chapters may inform the decisions that academic leaders, funders, other interested stakeholders, and both current and aspiring academics will continue to have to make over the course of the COVID-19 pandemic. Therefore, the committee identified, named, and documented preliminary evidence available to them to provide a framework that might help academia be attentive to the yet-unknown long-term impacts of the COVID-19 pandemic, especially on the academic careers of women in STEMM.

Beyond the COVID-19 pandemic, confounding and unexpected issues such as racial unrest and attention to natural disasters have already spurred

---

[10] This series of webinars comprised the "Workshop" referenced in the Statement of Task.

departmental committees, special councils, and task forces, and have identified new training needs, all of which will take time and require examination. By the end of 2020, preliminary evidence indicated that it is possible for leaders within institutions to learn from the COVID-19 pandemic experience. These lessons may include how to investigate acute issues that require immediate response and connect layers or aspects of their institutional infrastructure that may have been barriers to the retention and advancement of women in STEMM academic careers all along. In the same way that the *Promising Practices* report guided this report's examination of immediate effects of the COVID-19 pandemic on women in academic careers, identifying key questions about the longer-term effects of the COVID-19 pandemic can create a research agenda that will better prepare higher education institutions to respond to disruptions and explore opportunities in the future in ways that support the full participation of women. These are questions that must be asked, investigated, and answered with an equity lens.

Over the longer term, there are two certainties. First, there will be future disruptions—local, national, global, environmental, social, and political—that will test the principles and resilience of institutions of higher education. Second, STEMM fields will contribute to society to their maximum extent only if the well-being of women in these fields does not suffer significantly from these disruptions.

2

October 2020 Women in STEMM Faculty Survey on Work-Life Effects of the COVID-19 Pandemic[1]

COVID-19 has prevented me from working face to face with students and colleagues, traveling for work, and working in the lab, all of which are critical to my work as an experimental astrophysicist.
– STEMM woman faculty member (rank unknown)

Because I work from home, I have to hole up in my bedroom for work meetings, and because my husband and I both work full time, jobs that require meetings with other people, we constantly have to switch back and forth between roles. I get an hour or two for some Zoom meetings, then it's my turn to play kindergarten teacher for two hours, then I might get another hour or two to work. The constant task switching is mentally challenging and makes it hard to dive deep into any work task or accomplish anything that requires sustained attention for a longer period of time.... There are no boundaries between personal and professional life anymore. I really miss going to my office for many reasons, but being able to compartmentalize work and home ....is one of them.
–STEMM woman associate professor

---

[1] This chapter is based on the commissioned paper "Boundaryless Work: The Impact of COVID-19 on Work-Life Boundary Management, Integration, and Gendered Divisions of Labor for Academic Women in STEMM," by Ellen Ernst Kossek, Tammy D. Allen, Tracy L. Dumas.

INTRODUCTION

The biggest challenge the committee faced in trying to identify positive and negative effects of the COVID-19 pandemic on women in academic science, technology, engineering, mathematics, and medicine (STEMM) for this report was the lack of data, which was not surprising given that the committee was engaged in its work as the COVID-19 pandemic was still raging. This chapter includes findings from a national faculty survey conducted in October 2020 to examine the effects of the COVID-19 pandemic on women STEMM academics and that was conducted to gather original and timely data on post-COVID-19 pandemic work-life boundary and domestic labor issues specific to women in STEMM.[2] This survey asked women faculty in STEMM fields to compare how the COVID-19 pandemic has affected them between March 2020 and October 2020.

PREFERENCES AND CHANGES IN NUMBER OF DAYS WORKING AT HOME

The survey asked respondents to indicate their preferred and actual number of days working at home or on campus, both pre- and post-COVID-19 pandemic. Regardless of family or personal demographics, tenure status, rank, or caregiving responsibilities, women faculty reported working at home an average of 3.9 days compared with 0.66 days before the COVID-19 pandemic. In general, the preferred number of days working at home was significantly lower than the current number of days working at home, across all respondents.

Women faculty with children reported spending more time working at home than faculty without children at home.[3] Although women faculty with childcare responsibilities worked at home significantly less than their counterparts before the COVID-19 pandemic, they worked at home significantly more than their counterparts without children once the COVID-19 pandemic began. For many working parents, this was more often than they preferred.

Changes in Boundary Control

Boundary control, defined as the ability to control the permeability and flexibility in time, location, and workload between work and nonwork roles to align with identities (Kossek et al., 2012), is linked to important outcomes, such as work-family conflict and job turnover (work-life boundaries are covered more

---

[2] Details about the survey methodology are available in Appendix A.

[3] While the vast majority of institutions closed in-person operations due to state or local mandates at some point during 2020, the designation of faculty as "essential workers," requirements for remote work, and both online teaching and in-person teaching expectations varied, as did the duration of the closures.

fully in Chapter 4). Across the sample of women STEMM faculty, all reported significantly lower levels of boundary control after the COVID-19 pandemic began than before, and women faculty with childcare responsibilities reported significantly lower levels of boundary control during the COVID-19 pandemic than their women counterparts without children, including those with eldercare responsibilities.[4]

Survey respondents echoed these quantitative changes in their comments collected as part of the qualitative portion of the survey. Some 25 percent of the respondents wrote about experiencing a lack of boundary control when it came to preventing interruptions, particularly when scheduling teaching or managing virtual videoconference meetings. One example of this related to the inability to control family boundaries interrupting work demands during synchronous teaching. As a full professor, married with children, bemoaned, "… my son had a meltdown 5 minutes before my Zoom class was supposed to start." Another example of difficulties in managing boundary control involved trying to work and multitask while caring for children, which women ended up doing more often than their men spouses. One associate faculty member stated the following:

> *I teach synchronously via Zoom. My husband is home and does the same thing. He and I have some classes that overlap, which means that I must frequently teach with my daughter in the room with me. She's too young to understand how much I need her to play independently during class time, and I have lost a lot of a sense of professionalism with my students, because they see me getting constantly interrupted with comments like "Mommy, I went poop!!"*

Changes in Blurring of Work-Life Boundaries

Slightly more than half of the respondents mentioned having problems managing boundaries between work and personal life since the COVID-19 pandemic began. More than one-third of total respondents reported they were experiencing high boundary permeability as mentally and cognitively stressful to regulate.[5] Examples included the following:

- Blurred boundaries between work and family roles. For example, an assistant professor who is married and managing childcare comments, "They are happening simultaneously. I am working, and I am caring for my

---

[4] Details about the analysis used for the survey results are available in Appendix A.
[5] For an example mirroring this issue in academic medicine, see: Strong et al. (2013).

1-year-old. I am answering emails while making dinner. I am recording lectures while he naps. There are no boundaries, as everything happens at the same time and in the same space."
- A lack of time buffers between role transitions. For example, a married assistant professor without children stated, "Working from home, I log in and start looking at emails and responding to questions soon after waking up. The personal time that was earlier needed to get ready and commute to work provided the much-needed buffer between work and daily activities."
- Difficulties detaching from work when working at home. For example, an assistant married professor without care demands noted, "I'm always at home. Everything occurs at home. It's harder to turn off at the end of the day because there is no longer an end of the day."
- Limited space to create physical boundaries, including space for a separate office. For example, an assistant married professor with no children stated, "I don't have a closed office at home, since we can't afford a place that large. My husband has to work from home even without the pandemic, so he gets the one spare room. This means I have more distractions, kitchen noise, road noise, and a spouse who keeps walking into my 'office' at all hours. It is manageable, but psychologically is harder for me to keep the lines blurred, especially since I am just in my living room."

EFFECTS OF THE COVID-19 PANDEMIC ON WORK PRODUCTIVITY, WELL-BEING, CHILDCARE AND HOUSEHOLD LABOR, AND ELDERCARE

Effects on Work Productivity

The survey asked respondents how the COVID-19 pandemic has affected their personal and professional work outcomes, with the results summarized in Figure 2-1. Almost three-quarters of the participants mentioned the negative impact of the COVID-19 pandemic on their work. The most commonly mentioned top negative effects were the following:

- Increased workload resulting from more meetings, longer hours, more emails, and need for extended availability (27 percent). As one partnered faculty member without children commented, "I feel like my workload has increased by 50 percent. I'm not able to keep up. I am worn out and tired of having to constantly apologize for being late."
- Decreased work effectiveness (25 percent), with examples including decreased productivity, decreased efficiency, always being behind schedule, having tasks take much longer, and finding it hard to focus.

Effects of the COVID-19 Pandemic on Work Effectiveness of Women in Academic STEMM (N = 763)

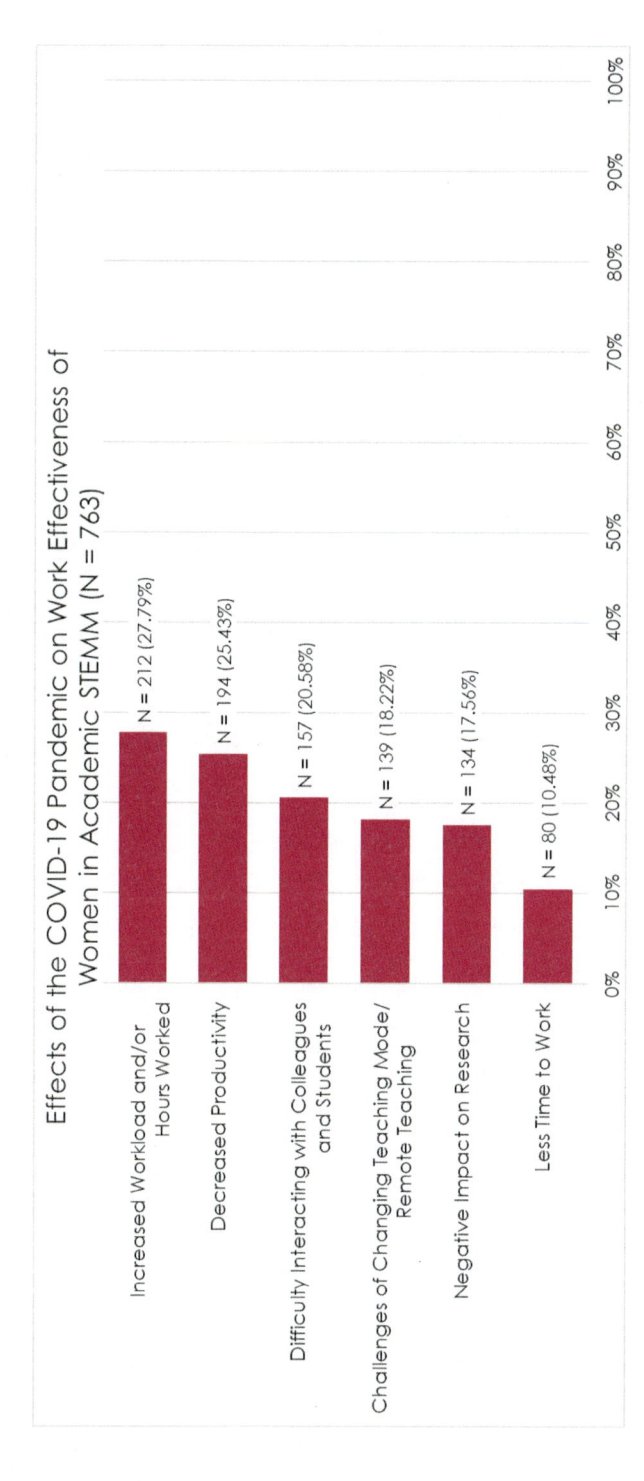

Increased Workload and/or Hours Worked	N = 212 (27.79%)
Decreased Productivity	N = 194 (25.43%)
Difficulty Interacting with Colleagues and Students	N = 157 (20.58%)
Challenges of Changing Teaching Mode/Remote Teaching	N = 139 (18.22%)
Negative Impact on Research	N = 134 (17.56%)
Less Time to Work	N = 80 (10.48%)

FIGURE 2-1 Summary of effects of COVID-19 on the work effectiveness and productivity of women in academic STEMM from the October 2020 survey.

- Poorer social interactions with peers and students (20 percent).
- Adverse effects on teaching and research (20 percent).

Other concerns were not having enough time to work and decreasing resources and support, such as pay cuts, furloughs, worries about research funding, and tenure outcomes and delays. (These topics are covered more fully in Chapter 3.)

Effects on Personal Well-being

Two-thirds of the respondents reported a negative impact on personal well-being, and 25 percent reported a decline in psychological well-being, regardless of rank and personal demographics. A married full professor with no children commented, "[I have] enormous stress—from work, family, trying to figure out how to work remotely… coping with an ever-changing array of rules, protocols, scenarios, problems." Similarly an assistant professor who is living with a partner stated, "There's a major increase in stress and anxiety as I feel like I'm working more/harder and accomplish[ing] less. This stress has taken a serious toll on my personal well-being." More than 6 percent of the respondents said the COVID-19 pandemic was causing a lack of sleep. One assistant professor with children described being "constantly stressed that the lack of lab productivity will cause me to not get tenure. I lose sleep over it." (See Chapter 7 for more information on these topics.)

Effects on Childcare

Figure 2-2 summarizes the challenges and coping strategies related to childcare demands reported in the survey. Nearly three-fourths of responding faculty with children reported a negative effect from increased childcare demands. A key reason for this is that 90 percent of women faculty were handling a majority of school and childcare demands. Only 9 percent of women reported that they shared childcare demands equally with their spouse, and only 3 percent said they had help from a babysitter, nanny, or tutor during the COVID-19 pandemic. Approximately 10 percent of the sample reported being the primary caregiver for children in their homes even if they were married to another professional. The effects of the pandemic were not all negative, as approximately 13 percent of the women respondents mentioned positive effects of the COVID-19 pandemic on family life, such as enjoying more family time together, ease in managing work-family demands, not having to dress for work, and a shorter commute.

Challenges and Coping Strategies of Women in Academic STEMM Related to Childcare Demands due to the COVID-19 Pandemic (N = 444)

Challenges
- Increased Childcare Demands: N = 316 (71.17%)
- Heightened Behavioral and Academic Needs: N = 245 (55.18%)
- The Lack of Childcare Feasibility, Accessibility, and Affordability: N = 213 (47.97%)

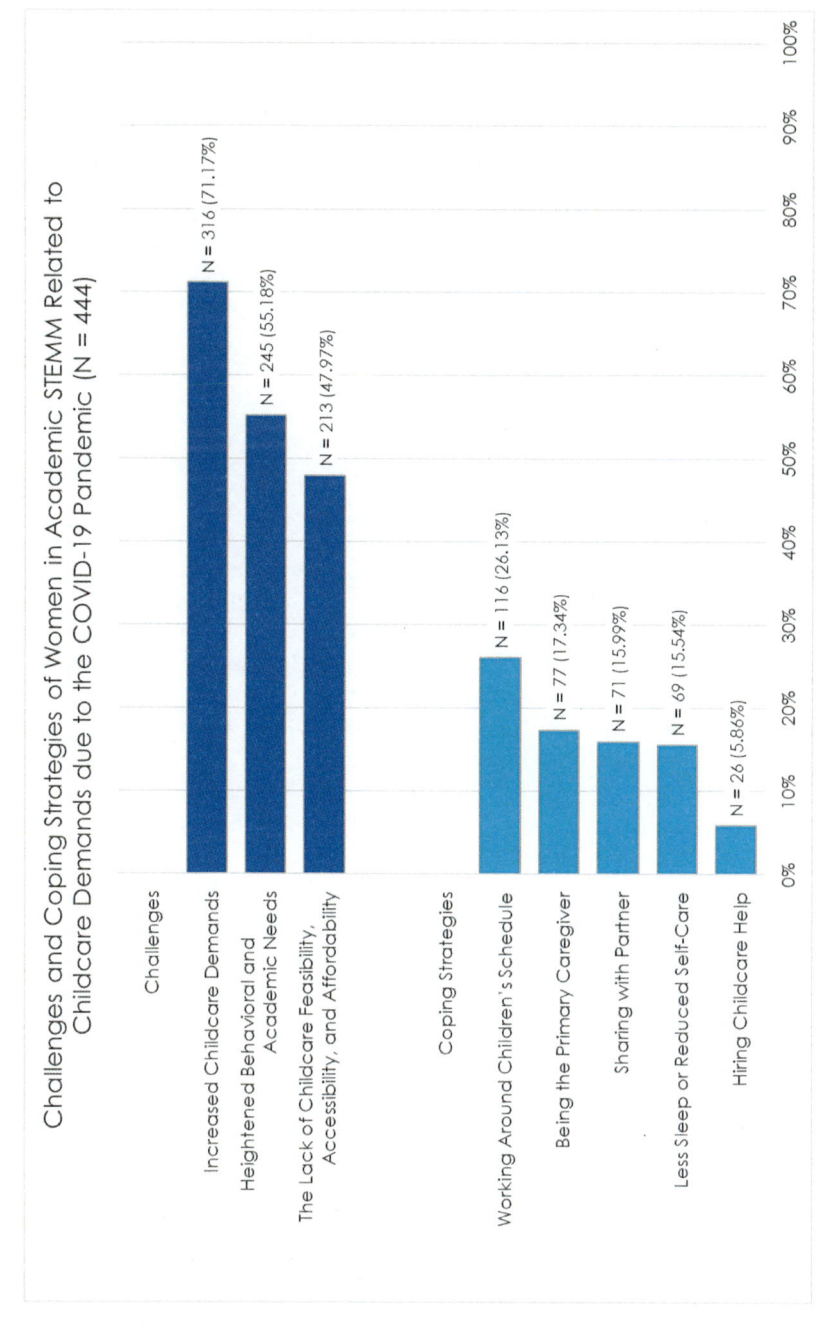

Coping Strategies
- Working Around Children's Schedule: N = 116 (26.13%)
- Being the Primary Caregiver: N = 77 (17.34%)
- Sharing with Partner: N = 71 (15.99%)
- Less Sleep or Reduced Self-Care: N = 69 (15.54%)
- Hiring Childcare Help: N = 26 (5.86%)

FIGURE 2-2 Challenges and coping strategies related to childcare demands reported in the October 2020 survey.

Childcare Feasibility, Accessibility, and Affordability

As a result of the COVID-19 pandemic, many respondents reported avoiding outside childcare because they were concerned about viral spread in childcare facilities. Some faculty reported an increased financial burden resulting from childcare needs, and some women reported they continued to pay for spots in childcare centers to prevent from losing their spaces while shouldering home schooling and childcare responsibilities themselves. A married assistant professor with young children shared the following:

> *We are trying to stay in our bubble, so we don't have any childcare for our two kids. We don't want to bring in babysitters or have day care unless absolutely necessary. But this means the kids are with us all the time except about 10 hours of in-person school a week.*

Those who desired to use outside childcare reported difficulty finding it. Examples from the survey include the following:

> *This is bonkers. I cannot find childcare for my youngest (three years old) and my older two children are remote learning for kindergarten and second grade. Babysitters/nannies in this area have raised their prices and now the starting rate is $20/hour and for three kids with remote learning duties have been offering $30-40/hour and still have not found someone to help. So since March, my husband and I have been simultaneously performing parenting full time and working full time. It is fundamentally exhausting.*
> – Associate professor, married with children

> *My husband and I are both pre-tenure faculty and we have two young children at home. We are both trying to maintain jobs that want to demand 150 percent of our time when we are having to split shifts (two hours in the home office then swap and two hours with the kids).*
> – Assistant professor

> *My husband and I both work full time jobs, remotely. We live in an 800-square-foot apartment in XXX. My 5-year-old is doing blended learning. We have to maintain our jobs and step in as kindergarten teachers (for a kid who absolutely does not want to do remote school work). There is so much more work to do to care for our kids (we lost our hired caregiver when the pandemic started) and only one adult can do work for their paid job at a time because the other has to watch our kids. In the spring, when this started, I had to stay up working very late into the night every single night to just barely keep my head above water and stay on top of my work.*
> – Associate professor

Home Schooling and Increased Household Labor Stress

More than 41 percent of the respondents reported that home schooling increased their workload and stress. More than one-third of the entire sample, regardless of caregiving demands or relationship status, reported strain from increased cooking, cleaning, and other domestic demands. One associate professor described the following:

> *I am on the verge of a breakdown. I have three children doing virtual schooling full-time who need my attention throughout the day; they all have different break schedules and seemingly interrupt me every 10 minutes. I want them to learn and thrive and I try to make these difficult circumstances for them as positive as possible, which means giving more of myself and my time to them. I try to wake up before them and work after they sleep, but this is hard given they wake up at 7 AM for school and don't go to bed early (they are 13, 11, and 8). There are sports/activities, dinner, homework/reading, etc. All the things that keep my evenings busy when they were in school, but now it is all day.*

Children's Heightened Behavioral and Academic Needs and Relational Strain

Some respondents indicated that their children across all ages, even into high school, were not adjusting well to remote learning and the disruption to their regular schedules. Therefore, some children needed more academic assistance from their parents and others acted out, further disrupting the faculty members' ability to work. These challenges also put a strain on relations between children and parents, and children and spouses in the household.

> *As a professional engineer working in academia, and single mother of three girls, the pandemic has radically changed everything. Although I spend more time with my girls, their mental health has deteriorated significantly with online school and very minimal contact with friends. Our social bubble with one other family (kids same age and gender) has been the only outlet. Even if there were enough hours in the day, I simply do not have the mental bandwidth to be a full time homeschooling mom, housekeeper, instructor, researcher, and family member (maintaining my family relationships from a distance—parents, sister, etc.).*
> – Associate professor, single with children

> *Being able to focus, and constantly shifting schedules to deal with kids and my husband's job. My 7-year-old is struggling with being home all the time and having a baby at home. So on top of the scheduling challenges, she is having way more behavioral problems than normal, which makes it even harder to work.*
> – Assistant professor, married with young children

On the negative side, I have children in school attempting to do virtual learning; this has been very difficult to manage while still trying to work myself. I have had to spend anywhere between one and three hours per day managing their virtual school activities. My husband does not feel as obligated and does not perform these tasks related to checking their schoolwork. I have lost sleep trying to make up for these lost working hours after the kids are in bed.
– Assistant professor, married with children

My son, although fairly independent as a high school student, is not adjusting well to virtual learning. His grades do not at all indicate his understanding of the content of his classes. He is finding it difficult to understand what the teachers are looking for through his virtual interactions with them. This has produced the need for frequent difficult family conversations that did not exist pre-COVID.
– Full professor, married with children

I also feel like I'm being put in the role of a mean mommy telling them they have to work extra at the end of the school day because they didn't get their work done during the day. I know that if they were physically in the classroom, the teachers would see them not being focused and the teachers could be the one encouraging them to work more efficiently. I guess I'm concerned about how online schooling is impacting my relationship with my kids.
– Associate professor, married with children

Eldercare and Sandwiched Care

Some 56 percent of the respondents reported increased eldercare demands, and nearly a quarter of those with elderly relatives reported increased stress from not being able to visit them. Figure 2-3 provides a summary of the challenges and coping strategies related to eldercare demands reported in the survey. Responses generally reflected three issues: demands associated with moving the family member from their initial care facility either to another facility or to have their parents move in with them to avoid exposure to the COVID-19 pandemic; the need to provide increased domestic support, such as household cleaning or ordering groceries, to minimize their elder's risks to the COVID-19 pandemic or the loss of paid support to provide eldercare; and concern over distance from the family member for the family member's well-being. As one married assistant professor with both childcare and eldercare ("sandwiched care") responsibilities noted, "I need to shop, cook, and provide all support for healthcare visits for both parents, one who died unexpectedly in July and has left us grieving on top of all this. Now mom is at home alone and needs more support and love in the middle of all this."

Another associate professor faculty member noted she was constantly

Challenges and Coping Strategies of Women in Academic STEMM Related to Eldercare Demands due to the COVID-19 Pandemic (N = 79)

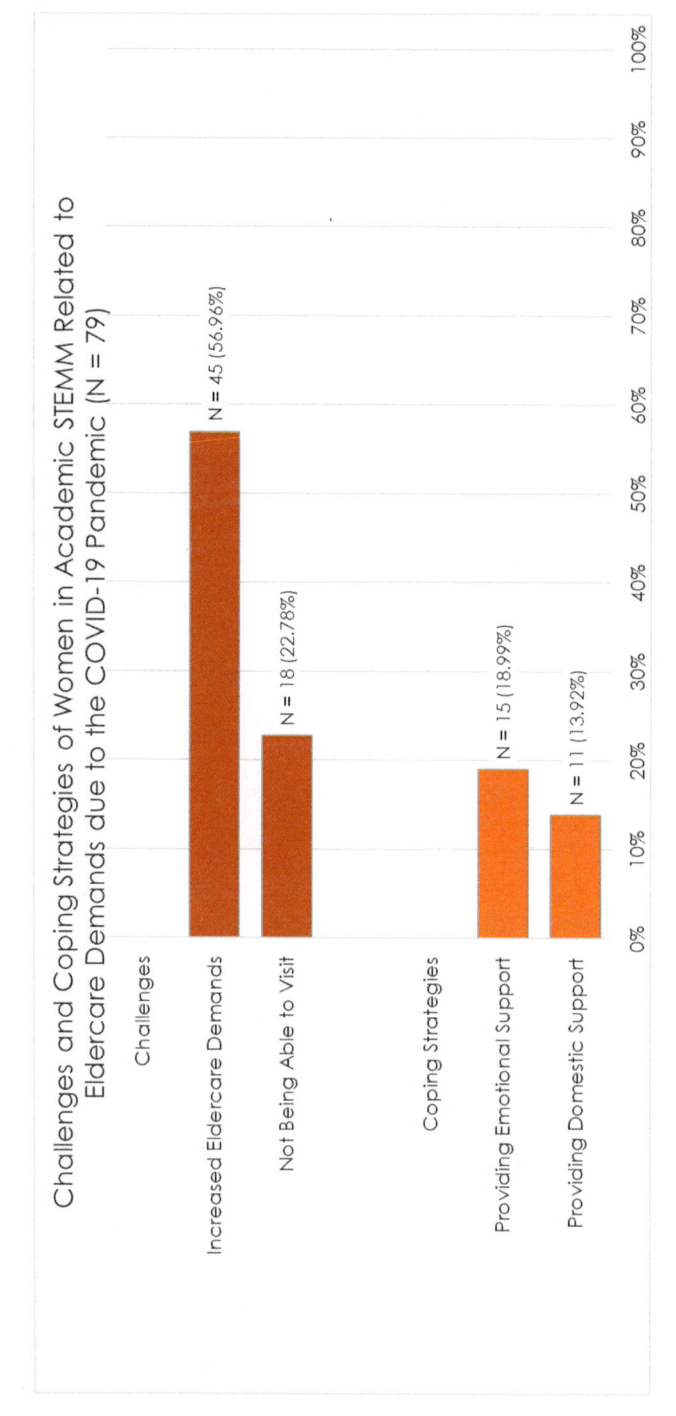

Challenges
- Increased Eldercare Demands: N = 45 (56.96%)
- Not Being Able to Visit: N = 18 (22.78%)

Coping Strategies
- Providing Emotional Support: N = 15 (18.99%)
- Providing Domestic Support: N = 11 (13.92%)

FIGURE 2-3 Challenges and coping strategies related to eldercare demands reported in the October 2020 survey.

stressed by the "inability to be able to fly back home to take care of [her parent] (or if bad things happened later). The anxiety of being stuck far away and not even knowing if I can attend the funeral on time is too high." A full professor who is unmarried reported that her parents are also exhibiting increased stress, resulting from "cancelled doctor appointments, more difficulty getting them care, multiple hospitalizations, move to facility, no visitation at facility, more mood disorder, isolation, unable to get services to home due to fear of COVID."

COPING STRATEGIES FOR BLURRED BOUNDARIES AND DOMESTIC LABOR

Many faculty actively used separation tactics to manage the boundary between work and home. Figure 2-4 summarizes boundary management tactics and other coping strategies that survey respondents have adopted during the COVID-19 pandemic. The most popular separation tactic involved the use of technology to hide the home space during videoconferencing, whereby faculty set up video meeting backgrounds to protect home privacy. The second most common separation strategy was having a separate office at home if space permits, such as turning a dining room into a designated office or setting up a computer in a guest room. However, some faculty respondents did not have a space that was conducive for work and sometimes needed to manage boundaries with family in ways children may not fully understand. A married assistant professor with young children shared the following:

> *I have a workspace set up in my walk-in closet, and I purchased a folding screen to put behind me so others can't see my dirty laundry or items all over the floor. I'll shut the door to the closet so the cats and kids stay out when I'm "at work." But if I need an extra level, the closet door leads into the bathroom so I'll close the bathroom door, which has a lock on it for extra assurance. However, my 5-year-old has figured out how to stick a bobby pin into the knob to pop the lock open when she's desperate.*

Respondents also used email boundary management as a means of limiting availability for work communications during nonwork times, with some faculty members physically removing or turning off work emails from their smartphones. Other respondents used one device for work and another for personal use. Some of these physical technology boundaries were combined with spatial boundaries.

Individual Temporal Strategies to Manage Working Time

Since most academic institutions did not have any policies or approaches in place to create a culture that helps faculty avoid overworking, many faculty had to self-regulate and triage new ways of coping with managing work and nonwork

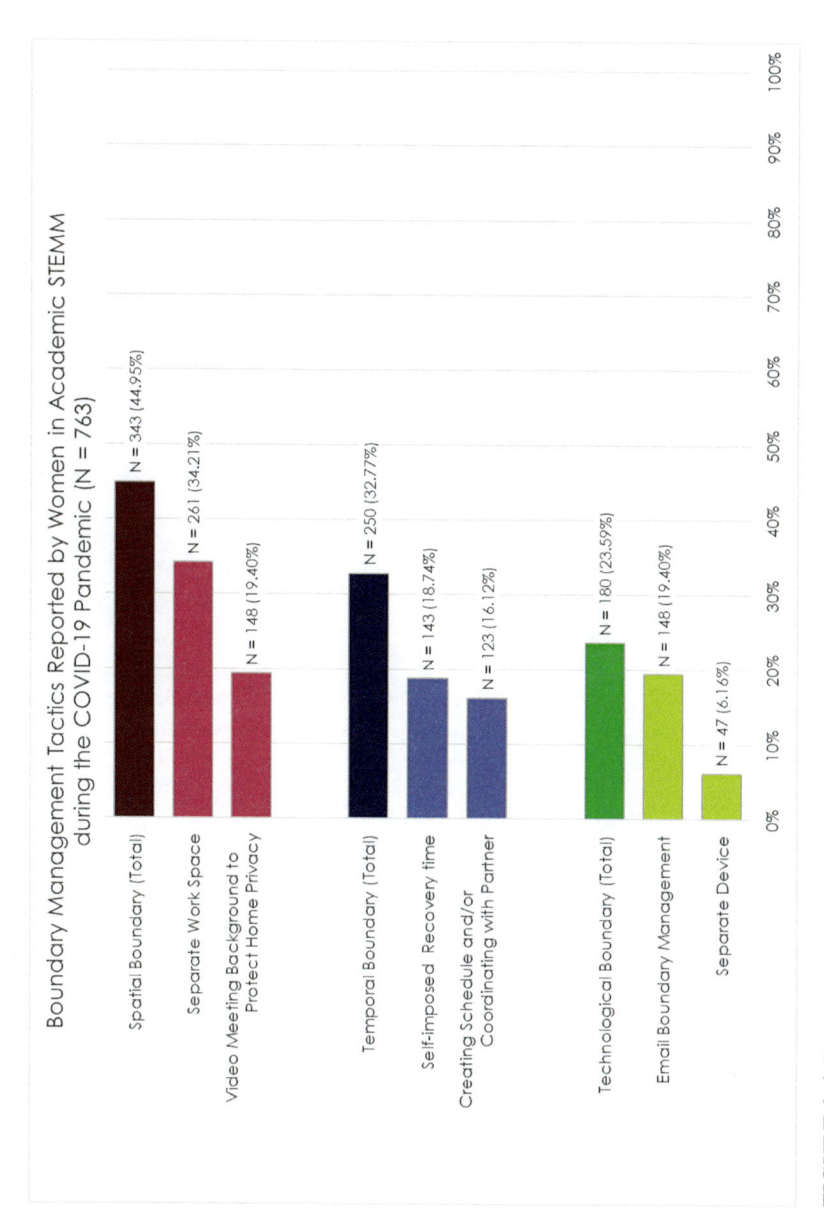

FIGURE 2-4 Boundary management tactics and other coping strategies reported in the October 2020 survey.

boundaries on their own, given their workload increased exponentially. Those with children in particular had to self-manage and engage in significant time restructuring to manage their heavier workload with childcare. The most common coping strategy, particularly for those with children, was working outside of standard hours and around children's schedules, which resulted in extended availability to work and long hours. Some respondents reported getting up early in the morning and working late at night when children were sleeping, and to a lesser degree, just sleeping less. As one associate professor with children commented, she and her spouse now "go 'back to work' after the children are in bed and it is still not enough time to keep on top of everything." Another assistant professor with children commented on the demands of juggling childcare/e-learning: "I can't get work done productively during the day, so work bleeds over until late evening. Regularly work from 9-midnight and start at 3 AM now."

Another temporal strategy for those with children involved setting up a coordinated work schedule with a partner, with periods of integration and separation to cover shared caregiving. Others organized their household with shared calendars with a spouse, if married, or blocked out time from work to take care of their children's schooling needs. An assistant professor with children commented:

My children have one remote learning day a week in their K–12 public school. I blocked off this time on my work calendar as a private appointment. I wanted to keep this time free to be available to help my children. As they settle into their remote learning routine, I find that I can work next to them. I am so glad I thought ahead to block off this time so that I am not torn between sitting with my children or being in another room occupied with a video meeting.

Some faculty set aside weekends to take off from work and allow for recovery, though enforcing this break was not always easy to accomplish.

Reducing Time Allocated for Self-Care

Given limited time, many faculty are putting their family's well-being ahead of their own. Many mentioned they have no time for themselves, as well as a lack of social support. While some respondents did mention self-care strategies such as taking walks, exercising, and meditating, many reported unhealthy strategies. As one married assistant professor with children explained, "I have had to reduce my sleep to a bare minimum (2–3 hours), forgo exercise or time to myself, and endure significant stress and anxiety."

ACTUAL VERSUS DESIRED UNIVERSITY ACCOMMODATIONS POST-COVID-19 PANDEMIC

Faculty respondents reported three main ways that academic institutions helped manage challenges associated with the COVID-19 pandemic:

- Giving faculty the option to work remotely.
- Extending the tenure clock.
- Allowing faculty to choose their preferred mode of teaching, whether online, remote, hybrid, or face-to-face.[6]

Given the sudden, unprecedented onset of COVID-19 pandemic challenges, some faculty reported that their academic institutions focused on testing and health issues (McAuliff et al., 2020) but did not have a plan or clear policies in place to help faculty working remotely. For example, at some institutions there was no infrastructure for childcare, school for children, or ways to continue research or reduce teaching demands. Switching to online instruction dramatically increased faculty members' workload. For example, faculty members needed to learn new technologies and redesign entire classes for remote learning almost overnight, something that is likely to change higher education for decades (Alexander, 2020). Furthermore, because of variance in student internet access and schedule control from home, faculty needed to deliver content both synchronously and asynchronously, resulting in the need for additional measures, such as recording lectures and remaining available for student interaction outside of normal class time or office hours (Alexander, 2020).

Lack of Caregiving and School Support

When asked how their academic institutions could improve in their handling of the COVID-19 pandemic, some faculty stated their academic institutions could have done a much better job of providing childcare, help with their children's schooling, and financial support. Though these were the supports faculty wanted, few academic institutions provided them. Instead, most academic institutions took a hands-off approach. An assistant professor with children stated the following:

> *Many faculty were expected to manage childcare demands by themselves. We were told to have backup childcare this semester in case schools closed (they are virtual part time), but they haven't offered any options or financial support for this in a town where daycares already had a >12 month waiting list pre-COVID, and they stopped allowing kids on campus.*

---

[6] Although some academic institutions did not give faculty members any choice in teaching mode, many provided options for faculty who are in higher-risk brackets for COVID-19 to teach remotely or in person as well as resources for faculty who need to teach from home.

Moreover, the underinvestment of some academic institutions in childcare support became more apparent when the COVID-19 pandemic began. As one faculty member with children commented,

> *Our on-campus childcare situation is terrible, too little capacity and historically not high-quality care. With the onset of the pandemic, it was closed and some schools at the university stepped up and provided additional childcare subsidies to families who needed them, but it was not centralized or universal across the university. HR is now being entirely restructured so perhaps it will end up being more comprehensive, inclusive, and proactive. There is in general an utter lack of proactive care of people's needs.*

Finally, a handful of faculty commented that their academic institutions were not culturally supportive of family life. As one faculty member stated, "My university does not care about families. They don't even mention issues with childcare in messaging and blamed the lack of affordable childcare on 'community partners.' It has always been a problem here, which is probably why we have so few women as professors." Such comments suggest that maybe the COVID-19 pandemic could be a catalyst for institutions to reinvest in new solutions to foster gender equality (Malisch et al., 2020).

Workload Reduction

One suggestion raised by a few respondents was to reduce teaching and service demands for those with childcare and eldercare responsibilities, and to modify research expectations for tenure, given the COVID-19 pandemic. A number of respondents stated that they did not feel that a tenure clock extension was an effective means of reducing workload during the COVID-19 pandemic. Rather, those respondents noted that what they needed was an acknowledgment that these years will result in much lower productivity.

HIGHLIGHTS FROM NON-TENURE-TRACK FACULTY RESPONSES

Although the intent of the survey was to focus on tenure-track faculty members whose research was largely stopped by the COVID-19 pandemic, there were some insights gained from the responses of non-tenure-track faculty who are also facing difficult career challenges. Most of these faculty members were lecturers and clinical professors who bore the burden of heavy course revision to a virtual format.

Negative Work Effects on Non-Tenure-Track Women Faculty in STEMM

Similar to the women faculty who are tenured or on the tenure track, about three-fourths of non-tenure-track faculty mentioned the negative effect of the COVID-19 pandemic on their work productivity. The top two most mentioned negative impacts on work productivity were increased workload and decreased work effectiveness, which were similar to the same top concerns of tenure-track and tenured faculty. While the academic tenured and tenure-track faculty's third most common concern was on the negative effects of the COVID-19 pandemic on social interactions with peers and students, for non-tenure-track faculty the third most common concern mentioned was a negative effect on teaching. Key concerns included a large increase in workload and stress resulting from technology problems, having to offer multiple formats to students, developing new content, and a lack of clear directions from administrators on decisions that could help planning. The following are three sample illustrative comments highlighting these issues:

I feel like I am not as effective at instructing students as I was pre-pandemic or even during the quarantine period of the pandemic. Currently, with offering flexible solutions for students, I am pulled in too many directions and spend 2–3 times the amount of prep time on lectures and materials. Trying to deliver content to students in class AND online has been a tremendous challenge and I feel like I waste about 20 minutes out of every 75-minute lecture just trying to get the technology to work properly. I'm working at least 12 hours a day either developing materials for both types of instruction or trying to get caught up on grading assignments and providing adequate feedback to students. Even my weekends are now rarely my own, since this is the only time I can record content for some of my courses.
– Senior instructor, married with children

Added much more STRESS to life. Working more hours at home than I would ordinarily put into my day when I went to campus. Had to learn technology quickly and by myself for the most part (adult children were helpful, too). Spring I tried asynchronous instruction which was a LOT of work and students were not pleased at all. Changed to synchronous instruction in summer and currently and overall a much more pleasant and satisfying solution to the problem. Had to figure out on-line labs in spring which was a total disaster and most unsatisfactory for both me and the students. As a program, we did not offer labs in summer until we were able to meet face to face beginning in July. Labs are face to face this fall so only issues are that some students are quarantined and miss at least two labs minimum, and yet must be counted as "excused."
– Senior lecturer, married without caregiving responsibilities

Early in the pandemic (March and April), there was so much communication (much of it contradictory) from department, college, and university level admin that we were jerked every which way almost every day. Admin seemed to think you could totally redesign your course on a dime in the middle of the semester, and sent us ads from third-party vendors, as well as constantly changing policy edicts and requests for information. This pushed me to work 10–12 hours per day, seven days per week, and resulted also in very unhappy students. The stress was unbearable, and by June I was in ICU with a stroke. Thankfully I have recovered sufficiently to keep working. But I fault the university for the amount of stress they caused.
– Anonymous, married with eldercare

Less than 5 percent of women faculty who are not on the tenure track mentioned worrying about job security because their job is dependent on contract renewal or funding. Examples of the concerns that were expressed included the following:

Our institution is facing mandatory 10 percent budget reductions. I am in a vulnerable position as a nontenured academic lecturer (despite 25+ years' experience at this institution, women faculty member in STEM field). So, who knows? I try to be grateful I have a job, a job I enjoy, and I am healthy.
– Senior lecturer, married without caregiving responsibilities

There is no guarantee whether I can have a postdoc in the next six months because it all depends on my supervisor and the funding agency. There is no fallback in these times of pandemic.
– Postdoc, married with children

Non-Tenure-Track Faculty Desired University COVID-19 Pandemic Organizational Supports

In general, the views of tenure-track and non-tenure-track faculty were similar regarding how academic institutions were helping women faculty manage challenges associated with the COVID-19 pandemic and how their academic institutions could improve. However, non-tenure-track women faculty noted that some of the accommodations their academic institutions are offering, such as extending the tenure clock, simply did not apply to them because of their employment status.

I'm not feeling my institution is encouraging work-life integration as a whole. My immediate supervisor is very supportive of my decision to work exclusively from home. A few "atta-boys" are tossed by the Provost to thank faculty for their flexibility with coping with challenging times, but no real differences implemented EXCEPT allowance to take 2 personal days this

semester. That's nice BUT the semester is already one week longer than in the past. And, if you teach every day, which day am I to take off??
– Senior lecturer, married without caregiving responsibilities

In reality, the flexible work schedules, reduced schedules, job-sharing and alternate work duties options they offer simply do not apply to teaching faculty, especially those that rely on their income to support their family.
– Assistant professor of practice, married without caregiving responsibilities

While there was no consensus on the further practices academic institutions could adopt to help non-tenure-track faculty beyond the same workload reduction and childcare recommendations that some tenure-track faculty wanted, it appears that extra teaching support for grading and technology support for converting courses to virtual formats might allow non-tenure-track faculty to have respite from their higher teaching loads.

CONCLUSIONS

Between March and November 2020, there was little guidance regarding institutional policies—both structural and cultural—that indicated what will be most helpful (see Chapter 6 for more on this topic). Tenure clock extensions were widely implemented as policies to address the COVID-19 pandemic productivity challenges (see Chapter 3 for more on this topic). However, these policies were implemented without addressing the disparities in increased caregiving and job-related workload that women faculty across all ranks and job status are facing. Previous research reported that gender-neutral tenure clock stop policies reduce women's tenure rates while increasing men's tenure rates (Antecol et al., 2018). For tenure-track faculty, this means that tenure clock extensions may not have a positive effect on women's careers, and may have an adverse effect on women's tenure achievement and the retention of women faculty. Based on the findings of the survey, it is clear that some faculty believe that tenure clock extensions alone will not be sufficient to help pre-tenure faculty manage the negative career effects of the COVID-19 pandemic.

This survey aligns with others that have been conducted during 2020. For example, in a survey of 608 scientists, approximately one-third of nontenured assistant professors were dissatisfied with their work-life balance during COVID-19, while 26.5 percent of associate professors and 10.8 percent of full-time professors expressed dissatisfaction (Aubry et al., 2020; Wallheimer, 2020). Two-thirds of respondents believed that temporarily stopping the tenure and promotion clock would be helpful (Aubry et al., 2020; Wallheimer, 2020).

The survey results presented here also highlighted how the COVID-19 pandemic is affecting multiple aspects of employees' lives, including outcomes related to personal well-being and those of their children and partners. Showing organizations and teams how to respect others' management of work-life boundaries (see Chapter 4) and to preserve others' needs for boundary control may help prevent burnout. While the mental health of most employees has been taxed during the COVID-19 pandemic, it is affecting women disproportionately compared to men (see Chapter 7). Work support interventions, such as greater administrative help in managing the added demands associated with learning new technology platforms for online teaching, are critical in reducing stress. These types of interventions may dovetail well with family support interventions that help faculty with managing schooling and childcare demands.

3

Academic Productivity and Institutional Responses[1]

INTRODUCTION

This chapter focuses on the potential direct and indirect effects that the COVID-19 pandemic may have on career trajectories of women in the science, technology, engineering, mathematics, and medicine (STEMM) academic workforce.[2] The COVID-19 pandemic's potential influences on women's careers will likely affect distinct aspects of academic work, including research, teaching, and service. Women in various appointment types (e.g., tenure-track faculty, postdoctoral scholars) are expected to engage in these activities at different levels and toward different goals. For example, whereas the postdoctoral researcher is likely focused on building a research program that will position them for a future faculty position, a tenure-track faculty member's work profile will include a mix of teaching, research, and service and be oriented toward promotion to associate or full professor. It is critical, then, to account for these various appointment types to understand how the COVID-19 pandemic has the potential to affect women situated at different points of the academic trajectory.

Women, in fact, are distributed across the academic workforce in distinctive ways. One study, for example, found that white women, followed by Asian women, hold the vast majority of women's tenure-track appointments (Finkelstein et al., 2016). Meanwhile, Black, Indigenous, and Latina women tend to be concentrated

---

[1] This chapter is primarily based on the commissioned paper "The Impact of COVID-19 on Tenure Clocks, the Evaluation of Productivity, and Academic STEMM Career Trajectories," by Felicia A. Jefferson, Matthew T. Hora, Sabrina L. Pickens, and Hal Salzman.

[2] For this chapter, the STEMM academic workforce includes tenure-track and non-tenure-track faculty, postdoctoral researchers, and graduate students.

among the non-tenure-track appointments, especially in the part-time ranks (Banasik and Dean, 2015; Wilkerson, 2020). With Black, Indigenous, and Latina women situated in the most vulnerable ranks of the academic work force, it is critical to acknowledge that the COVID-19 pandemic is not affecting all populations in the same way. Rates of infection and death are greatest among Black, Indigenous, and Latinx communities, with Black people dying at 2.1 times the rate of white people (CDC, 2020a) and People of Color representing 78 percent of deaths among people under the age of 21 (Bixler, 2020). Given the disparate rates of documented cases and death caused by the COVID-19 pandemic, it is important to recognize that academics who are Black, Latinx, Indigenous, and other People of Color, especially those women situated in non-tenure-eligible appointments, may be experiencing additional stressors with likely effects on their productivity and outcomes.

To the extent possible, this chapter also considers how women academics situated in various types of institutions are experiencing the effects of the COVID-19 pandemic. Just as appointment type shapes an academic's work portfolio, so does institutional type. For example, Minority-Serving Institutions (MSIs) champion broad access missions, which means that they enroll large numbers of Students of Color, first-generation college students, and students from economically vulnerable backgrounds (NASEM, 2019a); these student characteristics play a role in shaping faculty work experiences and expectations. Academics working in small liberal arts colleges are likely to have experienced the teaching-related effects of the COVID-19 pandemic more than a postdoctoral researcher working in a research university laboratory, for example.

BROADER LABOR MARKET EFFECTS OF THE COVID-19 PANDEMIC

To contextualize how COVID-19 has the potential to affect women's academic careers and trajectories, it is helpful to consider the pandemic's overall gendered labor market effects. Recent economic recessions in the United States resulted in employment losses that were larger for men than women (Hoynes et al., 2012), but the recession triggered by the COVID-19 pandemic has affected women's employment more substantially than it has men's employment (Alon et al., 2020b). As of October 2020, the share of women who are working is the lowest since the mid-1980s, when labor force participation among women was lower but still rising (Kochhar, 2020). There are several reasons why this recession has hit women harder than men, but a major contributor is the higher concentration of women employed in low-wage jobs requiring face-to-face customer interactions (Allegretto and Cooper, 2014; Entmacher et al., 2014).

There is some optimism on the part of business leaders that the COVID-19 pandemic will lead to greater opportunities for women and People of Color to advance professionally; employees, however, are more skeptical. A survey conducted by Catalyst: Workplaces that Work for Women, in partnership with Edelman Intelligence, queried 1,100 U.S. adults in full-time employment between June 1 and 5, 2020, about their beliefs regarding the pandemic and gender equity in the workplace (Catalyst, 2020a). The respondents included 250 business leaders of large companies and 850 employees of large multinational companies. While the survey was not focused on women in academic STEMM fields, the data highlight relevant differences between employers and employees by gender. Women employees were somewhat more likely than men employees to express skepticism that their employer is fully committed to taking the action necessary to create a more inclusive work environment for women. Regardless of gender, more than twice as many employees, compared to business leaders, were more likely to report fear that the COVID-19 pandemic would negatively affect their prospects for promotion.

The ultimate effects of the COVID-19 pandemic on the working population, in light of shutdowns and furloughs, can be seen as a "career shock," which is defined in vocational psychology as a "disruptive and extraordinary event" that occurs outside of an individual's control and "triggers a deliberate thought process concerning one's career" (Akkermans et al., 2020). Studies of the effect of recessions on labor markets finds a "scarring" effect on careers—with diminished income and career advancement—even when delays in entering or losing employment result from structural factors in the economy rather than individual factors.

EFFECTS OF THE COVID-19 PANDEMIC ON ACADEMIC PRODUCTIVITY IN 2020

The COVID-19 pandemic affected the current and future academic workforce, which includes postdoctoral scholars, non-tenure-track faculty, and tenure-track faculty among others with teaching and researching responsibilities. At the end of summer 2020, the Bureau of Labor Statistics reported the largest-ever decline in college and university employment: "At no point since the bureau began keeping industry tallies in the late 1950s have colleges and universities ever shed so many employees at such an incredible rate" (Bauman, 2020). Moreover, because of declines in revenue from tuition, campus housing, athletics, and other sources, colleges and universities have instituted hiring freezes (UCB, 2020; UMR, 2020).

While many equate the notion of academic productivity with research-related products (e.g., publications, grants), academic productivity is much broader and varies by both appointment and institutional types. For example, faculty career success may involve taking on several types of demanding roles, including successfully writing and receiving large grants, running research laboratories, publishing in top journals, serving on many committees and engaging in public outreach, seeing patients if in clinical medicine, and mentoring and teaching large numbers of students (Kossek et al., 2019–2021). Non-tenure-track teaching faculty members are more likely to hold heavier teaching and service responsibilities. In the context of the COVID-19 pandemic, faculty members may be more concerned with their ability to provide high-quality instruction to students and to be evaluated for their instruction accordingly. Acknowledging that there are many forms of productivity, much of the emergent data concerning productivity available in 2020 focused on research productivity, as discussed below. In addition, the measures themselves may not reflect the relative value or importance of research and academic work (Bauerlein et al., 2010).[3] Longer-term effects of the COVID-19 pandemic could include the influence of changes in academic productivity on the career trajectories of STEMM researchers.

Effects of the COVID-19 Pandemic on Academic Productivity and Careers of Early-Career STEMM Researchers

Although the evidence as to how COVID-19 will impact opportunities in the academic STEMM workforce was still developing, studies in 2020 found that STEMM academic researchers, especially postdoctoral scholars, were deeply concerned about their future. For example, in a worldwide survey of 7,670 postdoctoral researchers in academia, approximately 61 percent of respondents believed their career options were negatively affected by the COVID-19 pandemic (Woolston, 2020c). Twenty-five percent of the respondents felt uncertainty regarding their future professional research careers. Entering the workforce during a recession, such as the one caused by the COVID-19 pandemic, may lead to higher rates of unemployment and lower wages compared with graduating during healthy or neutral economic times. Potential outcomes include a 2 percentage point higher unemployment rate (Rothstein, 2020); a 10 percent wage gap that took an average of 8 years to close (Oreopoulos et al., 2012); and other effects such as employment in smaller firms and lower occupational attainment at higher rates for individuals from lower socioeconomic status groups (Kahn, 2010; Kawaguchi and Kondo, 2019; Kondo, 2015; Schwandt, 2019). One group of researchers found "a large degree of heterogeneity in the costs of recessions" (Oreopoulos et al., 2012, p. 26).

Other postdoctoral researchers expressed uncertainty about their visas expiring, which has inhibited them from completing their research and publishing their findings (Woolston, 2020c). In addition, international STEMM scholars have voiced increased concerns about their inability to travel across borders as a result of COVID-19 pandemic-related travel restrictions (Woolston, 2020c). Respondents to the survey of postdoctoral scholars were also concerned that institutions would rescind offers as a result of the pandemic.

In addition to concerns about opportunity and stability of the academic workforce, early-career scholars have expressed concerns about competing in a

---

[3] It should also be noted that there are well-established disparities in academic publishing by gender (e.g., Raj et al., 2016) and race (e.g., Mendoza-Denton et al., 2017), as well as gender biases in citations.

job market given their reduced productivity. For example, some STEMM disciplines require field research or laboratory studies. Loss of data during entire field seasons, for example, could be catastrophic for long-term studies, as well as for completing a dissertation, postdoctoral project, or a scholar's early publications that can lead to academic jobs (Inouye et al., 2020). Given that disruptions in the early phases of a career can lead people to departing a profession and/or falling behind in terms of accomplishments and prospects for advancement, these effects on postdoctoral students and other early-career scientists should be a cause for concern among STEMM professionals, funding agencies, and postsecondary institutions (Shaw and Chew, 2020). Alternatively, those in mathematics, statistics, computer science, and economics had less of a decline in research time compared with those in experimental sciences (Myers et al., 2020). Trainees also used the COVID-19 pandemic as an opportunity to hone skillsets, attend virtual conferences, and participate in online learning; 72 percent of experimental science trainees and 50 percent of computation science trainees reported gaining from virtual learning during the pandemic's shutdowns (Korbel and Stegle, 2020).

Effects of the COVID-19 Pandemic on Academic Productivity and Careers of Women STEMM Scholars

During 2020, the COVID-19 pandemic appeared to compound longstanding challenges and barriers that have faced women and contributed to their continued underrepresentation in STEMM, such as the persistence of gender bias in peer review, hiring, and promotions, and of gender-based stereotypes (Beede et al., 2011; Ertl et al., 2017). This could affect the careers of women across academia and particularly in the STEMM fields (Hansen, 2020).

As discussed further in Chapter 4, women are more likely than men to be responsible for childcare at home, adding to the challenges associated with overcoming inequalities in the academic workplace (Bianchi et al., 2012; Oleschuk, 2020).[4] These challenges were reflected in measures of productivity that varied by discipline. Only 29 percent of computational life scientists and 10 percent of experimental life scientists reported more than 80 percent productivity (Korbel and Stegle, 2020), and those in experimental sciences across disciplines (e.g., biochemistry, biological sciences, chemistry, and chemical engineering) reported having 30 to 40 percent less time in research compared with prepandemic levels

---

[4] Even before the pandemic, far more women in STEMM fields left their professions (43 percent) compared with men (23 percent) after having their first child (Cech and Blair-Loy, 2019), and women generally shouldered more childcare and household responsibilities than men did (Jolly et al., 2014). As a result, scholars are calling on postsecondary institutions to provide childcare supports, increase funding opportunities, and carefully manage tenure and promotion criteria for women faculty (e.g., prioritize women-authored papers, monitor teaching and service responsibilities) (Cardel et al., 2020a, 2020b). See Chapter 4 for a detailed discussion on work-life boundaries and household labor before and during the COVID-19 pandemic.

(Myers et al., 2020). These reductions in research time were not tied strongly to career stage or facility closures but instead to gender and having young dependents (Myers et al., 2020). By occupying more experimental scientist positions, and accounting for childcare duties, women scientists reported even less productivity than men (Korbel and Stegle, 2020; Myers et al., 2020).

Early studies on academic publishing may also reflect differences in childcare responsibilities, with women's share of first and overall authorship in COVID-19 pandemic-related papers having decreased by 23 percent and 16 percent, respectively (Andersen et al., 2020), "whereas no significant differences in productivity were reported by men" (Krukowski et al., 2020). An analysis of all Elsevier journals also showed signs of gendered effects of the COVID-19 pandemic, especially in health- and medicine-focused journals (Squazzoni et al., 2020). In particular, the number of manuscripts submitted during the COVID-19 pandemic was significantly higher for men than for women compared with submissions in 2018 and 2019. In addition, researchers investigating gender disparities in published research during the COVID-19 pandemic found that, compared with models predicting authorship for women based on data from 2019, the proportion of women authors publishing on all topics as the first author decreased by 4.9 percent (Muric et al., 2020). The proportion of women writing on COVID-19 pandemic-related topics as the first authors dropped by 44.5 percent compared with baseline predictions of anticipated authorship among women, and when observing the authors regardless of the order, the proportion of women writing about the COVID-19 pandemic dropped by 15.4 percent compared with the baseline (Muric et al., 2020).

Other studies indicated that authorship and productivity might be dependent on discipline and field. While productivity and research time was shown to be heavily affected by gender and dependents (Myers et al., 2020), the American Geophysical Union (AGU) showed that the proportion of annual AGU journal submissions from women remained the same from 2018 to 2020, while the absolute number of year-to-date submissions increased from 2,632 submissions to 2,790 submissions from 2018 to 2020 (Wooden and Hanson, 2020). The study on the AGU journal submissions hypothesized that the increase in virtual collaboration encouraged larger teams and thereby resulted in increased team gender diversity.

Of particular concern is the effects of the COVID-19 pandemic on the careers and professional advancement of women in academic medicine, who have been forced to address a highly stressful and potentially dangerous workplace on top of added responsibilities at home (Madsen et al., 2020). When clinical workloads increase, faculty in academic medicine reduce their participation in scholarly research activities, such as publishing papers, writing grants, and completing research studies (Mullangi et al., 2020). In addition, women in academic medicine are more likely to have teaching-related roles instead of funded research positions (AAMC, 2016; Pololi, 2010), which translates into higher clinical workloads than many men bear. This disparity is likely a key aspect leading to the differential effects of the COVID-19 pandemic on women in academic medicine.

Effects of the COVID-19 Pandemic on Academic Productivity and Careers of STEMM Scholars of Color

While the specific effects of the pandemic on the careers of Scholars of Color had yet to be documented by the end of 2020, early signs indicate that preexisting inequalities are being exacerbated by negative effects on the financial situations of Scholars of Color and work-life balance (see Chapter 4 for further discussion of work-life balance). At least one author has noted that persistent disparities in funding for Black scientists by the National Institutes of Health (NIH), low numbers of Black women in science, and institutional racism were preexisting stressors that the COVID-19 pandemic and the civil unrest sparked by the killing of George Floyd only exacerbated (Carr, 2020). Addressing these disparities calls for institutional responses such as targeted recruitment and retention of Black women in STEMM fields, thereby creating an environment that encourages and nurtures diversity in collaboration and talent, a sense of belonging for Black women in STEMM, and positive morale for the institution's current and future researchers (Carr, 2020). Other authors have suggested that institutions of higher education need to more proactively support and invest in student-faculty mentoring relationships, access to professional development, and engagement in robust research opportunities for Women of Color (Ong et al., 2011).

Effects of the COVID-19 Pandemic on Teaching and Mentoring

During 2020, women faculty have reported having less time for advising, mentoring, and research because of their increased caregiving responsibilities (Anwer, 2020; Malisch et al., 2020). Specifically, women reported having had the same amount of course workload with the additional burden of transitioning their courses into a remote setting, while simultaneously experiencing the increased burden of dependent care. In another study, researchers interviewed 25 U.S. women faculty members and 55 Italian women faculty members, all of whom had children (Minello, 2020). These women reported reductions in their research productivity that was caused, in part, by the need to devote more attention to teaching online courses, which was difficult with small children in the home. Both real-time and asynchronous online teaching were interrupted by children's demands, cries, or other background noise. As a result, women faculty have less time for two other major aspects of their professional lives, mentoring and research (Kramer, 2020a; Minello, 2020; Zimmer, 2020).

According to an interim report on the COVID-19 pandemic and engineering education presented by American Society for Engineering Education, only 53 percent of faculty surveyed agreed they were given adequate resources from their institution to transition to online teaching. Most institutions did not reduce teaching load or advising and mentoring loads in response to the COVID-19 pandemic (Gruber et al., 2020). However, learning to teach remotely required more time from faculty (Alexander, 2020; ASEE, 2020).[5] It remains to be seen what the long-term effects of university professors transitioning to teaching hybrid online/in-person or entirely remote classes are or how these effects may be disaggregated using an intersectional framework.

Effects of the COVID-19 Pandemic and Institutional Context on Academic Productivity

A considerable body of empirical evidence exists demonstrating that resources, including federal research grants, are disproportionately allocated to high-prestige research universities (Taylor and Cantwell, 2018, 2019). In fact, approximately 50 percent of all STEMM research funding goes to about 100 doctoral-granting institutions, leaving the other 3,900 U.S. institutions—including community colleges, regional comprehensive universities, and MSIs—competing with one another for the remaining funds. Furthermore, researchers have documented considerable disparities in state and federal funding for Historically Black Colleges and Universities (HBCUs), with many state governments prioritizing predominantly white institutions (Boland and Gasman, 2014; Minor, 2008), leaving MSIs operating with less institutional support for STEMM research activities.

The effect of the COVID-19 pandemic on individual STEMM scholars likely depends not only on their race, gender, or disciplinary affiliation but also on the type of postsecondary institution where they work. Consequently, it can be hypothesized that a STEMM researcher at an HBCU, where scholars typically receive less research funding and have higher teaching loads than at predominantly white research universities, may have fewer opportunities to publish in the midst of a crisis such as the COVID-19 pandemic. However, future research will be required to isolate the effects of institutional affiliation on academic productivity during the pandemic, if such differences do exist.

EFFECTS OF INSTITUTIONAL RESPONSES TO THE COVID-19 PANDEMIC ON ACADEMIC CAREERS AND PRODUCTIVITY

Postsecondary institutions find themselves in uncharted territory financially as a result of the COVID-19 pandemic, and they have responded in several ways, including reducing overtime work hours for nonfaculty members, eliminating merit increases, reducing salaries of leadership members, obtaining additional

---

[5] Although some academic institutions did not give faculty members any choice in teaching mode, many provided options for faculty who are in higher-risk brackets for COVID-19 to teach remotely or in person as well as resources for faculty who need to teach from home.

governmental funding through the CARES Act (Coronavirus Aid, Relief, and Economic Security Act), and altering tenure and promotion policies. In addition to college and university changes in policies, many funders are allowing extensions on projects and other adjustments as necessary. A number of authors offered suggestions to institutions for how to address growing inequalities as a result of the COVID-19 pandemic (Cui et al., 2020; Guatimosim, 2020; Kibbe, 2020; Malisch et al., 2020; Oleschuk, 2020). One opinion piece, for example, offered 10 rules that women principal investigators could follow during the COVID-19 pandemic (Kreeger et al., 2020). These rules include a suggestion to find peer groups of women to provide support, saying no to nonessential responsibilities, dropping certain projects and tasks, and pushing back on demands to be more productive. Another opinion piece proposed solutions that include reevaluations of tenure-track and academic evaluation policies (Malisch et al., 2020). In reviewing the institutional responses to the challenges caused by the COVID-19 pandemic in 2020, it is important to consider not just the differential effects on women and People of Color, but also in the capacity of different institutions to respond and support women and Faculty of Color.

Funding Extensions

If principal investigators were unable to complete their research within the original time frame, many funders were more flexible and allowed no-cost extensions, revisions to the original budget, or even costed supplemental extensions in some cases (NIH, 2020b; UCL, 2020). For example, the NIH allowed clinicians to extend their research work if they postponed a career development award to aid frontline workers. In addition, NIH did not withdraw funding if there was a delay in starting a research project (NIH, 2020b). Although many funders modified their policies to allow greater flexibility to researchers in 2020, laboratory closures and other delays will extend the time of the research projects, and thus findings and publications, which may affect academic productivity. Moreover, even though funder policies allow for flexibility in project extensions, there was often no additional funding to support for staff and graduate students over the longer project period. No-cost extensions, in particular, often do not account for the salaries incurred by staff and trainees while they are unable to conduct research.

Promotion and Tenure Policies and Decisions

At least some institutions strove to account for the wide range of direct and indirect effects of the COVID-19 pandemic on academic productivity and careers. For example, a variety of colleges and universities (Chronicle Staff, 2020), such as Stanford University, University of Texas, and University of Washington, extended the tenure clock (Stanford University, 2020; UT System, 2020; UW,

2020).[6] At the University of Wisconsin–Madison, policies exist to extend tenure-track periods arising from extenuating circumstances, such as family health issues (Pribbenow et al., 2010). The University of Texas at Austin's updated policy states, "All tenure-track faculty are eligible to extend their probationary period for one year due to the negative impact of COVID-19."

However, extending the tenure clock tends to put off financial incentives and career advancement and freedom (Pettit, 2020a). As a result, while extensions during the COVID-19 pandemic may be important for some faculty members, they can also be unhelpful for career trajectories. Instead of extending the tenure clock, another approach could be to reward junior faculty for engaging in other avenues of research to which they could lend their expertise, such as studies related to the COVID-19 pandemic (Connolly, 2020). In addition, scholars of higher education have recommended that academic leaders be mindful of how the immediate transition to virtual teaching and learning may have taken a negative toll on faculty teaching evaluations and the potential subsequent effects on faculty promotions (Gonzales and Griffin, 2020). It has also been suggested that faculty document "how and what they learned while teaching through COVID-19" and detail the emotional effect they have experienced since caring for students to a different degree compared with before COVID-19 (Pettit, 2020a). While 1-year extensions and grant extension flexibility are helpful, overall, the differential effects for women may not be sufficient to address the added caregiver status and home responsibilities that affect work-life integration.

CONCLUSIONS

The effect of the COVID-19 pandemic on academic productivity and career trajectories cannot be adequately evaluated without acknowledging the intersecting identities of, and structural forces affecting, different groups of STEMM researchers. The effects of the COVID-19 pandemic on the academic STEMM job market, notions of academic productivity, and institutional responses each play out in different ways depending on the unique circumstances of individual institutions and individual STEMM researchers and faculty members. This is not solely an argument that "context matters" in dictating how national phenomena unfold in local settings, but a recognition that the individual lives of STEMM scholars and how they see themselves and their opportunities are deeply embedded in and shaped by these overlapping spheres of influence.

---

[6] An informally gathered list of changes to tenure clock policies is available at https://docs.google.com/spreadsheets/d/1U5REApf-t-76UXh8TKAGoLlwy8WIMfSSyqCJbb5u9lA/edit#gid=0&fvid=238051147.

4

Work-Life Boundaries and Gendered Divisions of Labor[1]

INTRODUCTION

Women academic scientists have long juggled unequal family caregiving and domestic demands and faced gender discrimination, and this is particularly true in science, technology, engineering, mathematics, and medicine (STEMM), where women are significantly underrepresented (Zimmer, 2020). While the COVID-19 pandemic is not responsible for the domestic labor challenges that increasingly affect the careers of many academic scientists, it has exacerbated them and shined a light on the work-life inequality that women experience, one that is a growing form of job inequality (Kossek and Lautsch, 2018; Kossek and Lee, 2020b). In fact, reports suggest that such work-life inequality could result in setbacks in gender representation and advancement in STEMM fields and the loss of early-career women in academics, particularly those with children (Cardel et al., 2020a). These reports indicate that growing numbers of professional women (Coury et al., 2020), particularly those in academia (Buckee et al., 2020), are considering cutting back or leaving the workplace altogether because of family demands brought on by the COVID-19 pandemic.

This chapter focuses on the ways in which the COVID-19 pandemic has affected the personal-professional boundary interface and work-life issues for women in academic STEMM; how gendered expectations of domestic labor and caregiving responsibilities for children and elders have shifted or affected professional labor and well-being for women; how research has informed emerging

---

[1] This chapter is primarily based on the commissioned paper "Boundaryless Work: The Impact of COVID-19 on Work-Life Boundary Management, Integration, and Gendered Divisions of Labor for Academic Women in STEMM," by Ellen Ernst Kossek, Tammy Allen, and Tracy L. Dumas.

individual boundary management and family-care coping strategies; and how the events of 2020 have widened the gap between current and desired organizational practices to support increasingly blurred work-life boundaries as well as preferences for integration and separation.

PRE-COVID-19 PANDEMIC WORK-LIFE LITERATURE OVERVIEW

This review of pre-COVID-19 pandemic literature is organized into two main parts, with a focus on data-based studies that were specific to academic women and particularly those in STEMM.[2] The first part provides a brief overview of work-life foundational concepts relevant to this report, including work-family (or personal) life conflict, enrichment, boundary management, and their relationships. The second part examines the implications of these concepts for women's careers in their academic social contexts, which have work structures and cultures that were largely developed before women increased their participation in STEMM fields. These themes reflect how work-family dynamics play out in academic social contexts that can be characterized as not being responsive to a growing mismatch between women faculty's career and personal life synthesis needs and the design of academic institutions.

Foundational Concepts from the Work-Life Literature

Work-Family Conflict, Enrichment, and Gender

Tensions between work and nonwork lives can be understood from the individual and organizational psychological science behind role theory and the associated concepts of role conflict and enrichment. All individuals have multiple roles in life—employee, parent, partner, daughter, and volunteer, for example (Katz and Kahn, 1966)—in which a role is defined as a position in a group or organization with accompanying responsibilities, rights, and behavioral expectations (Kahn et al., 1964). Role conflict occurs when an individual perceives incompatible time, strain, or behavior-based demands between work and nonwork roles (Greenhaus and Beutell, 1985; Kahn et al., 1964). For example, a tenure-track faculty member who is a parent may perceive that the behaviors she must carry out to care for her children interfere with the research, teaching, and service demands at an academic institution. Qualitative studies in academic medicine have described similar challenges (Strong et al., 2013). During the COVID-19 pandemic, it is likely that these work-nonwork demands may be increasingly at odds, particularly in families with school-age children when, for example, the

---

[2] The search terms that informed this literature review are provided in Appendix A.

adult is scheduled to teach a class via Zoom at the same time that a child needs help with online schooling.

Historically, work-family research has suggested that women's work-family experiences can differ from those of men. An early meta-analysis found that the relationships between work-family conflict and both job and life satisfaction had stronger negative associations for women than for men (Kossek and Ozeki, 1998). Evidence from another meta-analysis, conducted two decades later, suggests that as men become more involved in household tasks, they are starting to report as much work-family conflict as women do (Shockley et al., 2017). While men may believe they more equally share household tasks, the data on actual household labor time show that women with children under age 6 spend less time in the labor force and more time on household tasks than do men, a trend that continues for school-age children (Bianchi et al., 2012; Glynn, 2018) and generally for eldercare (Porter, 2017).

Complementing the literature on work-family conflict is a growing body of research on work-family enrichment, defined as the positive transfer of knowledge, skills, and emotions from one domain to another (Greenhaus and Powell, 2006). Work-family enrichment theory assumes that having multiple roles can benefit well-being. This relationship between multiple roles is most likely to occur when one's work and nonwork demands can be carried out in ways that align with preferences for how one synthesizes work and nonwork roles. For example, employed men reported positive work-to-family enrichment relationships in the transfer of positive emotions and engagement from the work-to-family realms, while women are depleted in the spillover from work-to-family roles (Rothbard, 2001). Though women also experience enrichment, it tends to go in the opposite direction, from the family role to the work role (Rothbard, 2001).

Boundary Management Strategies, Control, and Work-Family Conflict

Work-life boundary management is defined as the organization of work and nonwork roles to reinforce or weaken the boundary between them cognitively, physically, and emotionally (Allen et al., 2014; Ashforth et al., 2000; Kossek et al., 2012). Boundary control refers to an employee's ability to control how they manage the boundary between work and nonwork roles and considers whether an employee can maintain the boundary aligned with their preferences (Kossek et al., 2012; Wotschack et al., 2014). When individuals lack boundary control and the ability to choose the amount of work-nonwork segmentation, they have lower person-environment fit (Kreiner, 2006).

Individuals vary in the ways that they prefer to organize and synthesize work and nonwork roles to align with their career and family identities and roles (Kossek et al., 2012). Those who prefer integration are comfortable removing or blurring boundaries between work and nonwork, whereas those who prefer segmentation would rather keep boundaries between work and nonwork more intact (Allen et al., 2014; Ashforth et al., 2000; Kossek et al., 2012). Others cycle frequently through varying boundary styles as work- and family-role demands shift in peaks and valleys over time (Kossek, 2016). Figure 4-1 summarizes different boundary management styles validated in several studies.

Research suggests that an individual's preferred alignment of work and nonwork roles may shape their boundary management style and the degree to which they integrate and segment those roles (Kossek et al., 2012). However, besides family structures, organizational policies, job structures, and occupational norms may determine the extent to which individuals have the ability to integrate or segment work and nonwork roles, as well as their overall amount of control over the work-nonwork boundary (Allen et al., 2014; Ashforth et al., 2000; Kossek, 2016). Organizational contexts may also influence the degree to which one perceives the ability to access and customize work flexibility to manage boundaries and the effectiveness of boundary management strategies (Kossek and Lautsch, 2012; Kramer, 2020b; Rothbard et al., 2005).

In general, research shows that a more permeable work-nonwork boundary is associated with increased work-family conflict, increased distress, higher turnover intentions, and diminished work performance (Boswell et al., 2016; Chesley, 2005; Kossek et al., 2012). For example, interviews of Navy personnel, their commanding officers, and family members found that the use of cell phones and email while on duty resulted in distractions, interruptions, reduced productivity,

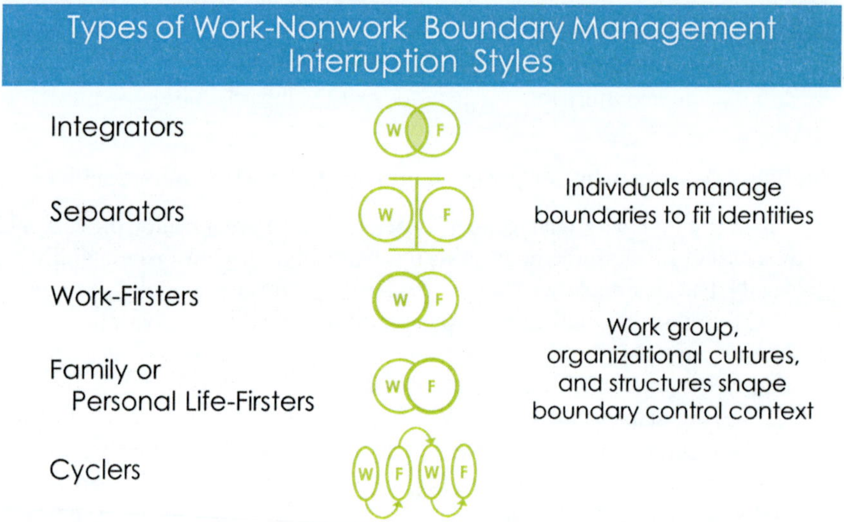

Types of Work-Nonwork Boundary Management Interruption Styles

- Integrators
- Separators
- Work-Firsters
- Family or Personal Life-Firsters
- Cyclers

Individuals manage boundaries to fit identities

Work group, organizational cultures, and structures shape boundary control context

FIGURE 4-1 Types of work-nonwork boundary management interruption styles.
NOTE: "W" represents work-related activities; "F" represents family-related activities.
SOURCE: Adapted from Kossek, E.E. *The Impact of COVID-19 on Boundary Management, Work/Life Integrations, and Domestic Labor for Women in STEMM*. Presentation to the Committee on November 9, 2020.

and mistakes at work, resulting in organizational policies restricting such work-nonwork integration (Stanko and Beckman, 2014). Permeable boundaries can make employees feel as if they never truly leave work behind, and they feel the burden of the expectation that they must be available at all times to meet work demands (Duxbury et al., 2014; Jostell and Hemlin, 2018). Such continuous availability is associated with increased work-family conflict (Eddleston et al., 2017; Lapierre et al., 2016), emotional exhaustion (Dettmers, 2017), and the inability to recover adequately from work (Dettmers et al., 2016).

For many professionals, including women in STEMM, creating separation between professional identities and personal boundaries can be challenging (Dumas and Sanchez-Burks, 2015). Studies show that work-life boundaries can be more permeable for women than men, as they are likely to interrupt work for family demands (Rothbard, 2001). As a result, variation in boundary management strategies can result in varying effects on work-family conflict and employee well-being, including outcomes such as engagement, stress, depressive symptoms, and exhaustion (Chesley, 2005; Olson-Buchanan and Boswell, 2006; Powell and Greenhaus, 2010; Rothbard, 2001).

Women's Second Shift at Work and Home, Diverse Needs, and Ideal Worker Tensions

The extra work and nonwork demands that women faculty face compared with their counterparts who are men are numerous. The term *second shift* is based on research showing that employed mothers face a double day of work (Hochschild, 1989). After returning home from a day of paid work, most begin their second shift of unpaid work that includes childcare and housework. Decades after second shift was coined, the gendered division of nonpaid labor remains (Shockley and Shen, 2016). Specific to faculty, time expenditure studies show that women faculty spend more time caring for children than do their men counterparts (Golden et al., 2011; Misra et al., 2012).

Eldercare, which also falls more heavily on women than men, has a different life cycle and care dynamics than childcare (Kossek et al., 2001). Though there are exceptions, such as in one study of faculty that found no gender differences in eldercare involvement (Misra et al., 2012), women account for more than 60 percent of caregivers for elderly parents or other aging family members (National Alliance for Caregiving and AARP, 2020). About 60 percent of eldercare providers work while caregiving, with most reporting that caregiving negatively affects their work (National Alliance for Caregiving and AARP, 2020). Those adults who care for dependent children and an older adult are referred to as "sandwiched caregivers." Women account for three in five sandwiched caregivers, who as a whole account for 28 percent of all caregivers (National Alliance for Caregiving, 2019). Sandwiched caregivers are more racially or ethnically diverse than non-sandwiched caregivers (Schiebinger et al., 2008).

Different family structures or marital status and household career configurations can privilege the caregiving resources available to men faculty who are more likely to be in family structures where their career is primary in a couple. For example, reports indicate that in dual-academic couples, men faculty are four times more likely to have a partner who provides full-time domestic care than are women faculty (Jolly et al., 2014). Similar findings have been reported among STEMM faculty.

Researchers have examined the time that physician recipients of a National Institutes of Health K08 or K23 award spend on parenting and domestic work (El-Alayli et al., 2018). Women in this study were more likely than men to have spouses or domestic partners who were employed full time. Moreover, among married or partnered physicians with children, women spent 8.5 more hours per week on domestic activities than did men after controlling for work hours and spouse employment (El-Alayli et al., 2018).

The gender differences associated with caring for others is not limited to home, and, in fact, women's care work roles often extend into the work domain. Women professors, for example, report having more teaching-related work and receiving more special favor requests from students than do men professors (Guarino and Borden, 2017) (see Chapter 3 for more on service tasks).

Academic Scientists as Overloaded Ideal Workers

With increasing workloads and the rise of personal electronic devices that blur work-life boundaries, many academic STEMM professionals face role overload. Similar to other professionals with a large investment in human capital, many STEMM faculty are socialized to work long hours after having invested years into earning a doctoral or medical degree and then working to advance in their careers to tenure and beyond. Such work devotion continuously competes with nonwork passions or interests (Blair-Loy and Cech, 2017).

Norms encouraging adherence to "ideal worker" behaviors contribute to the overwork pressures that often prioritize work over personal life (Kossek et al., forthcoming; NASEM, 2020). The concept of the ideal worker reflects a breadwinner-homemaker model that dates back to the Industrial Revolution (Williams, 2020) and perpetuates the myth that work and nonwork lives are "separate worlds" (Kanter, 1977). Ideal workers try to ensure that family or other nonwork matters do not hinder work commitments (Kossek et al., forthcoming). This behavior results in overworking, the idea of working more than is needed to perform one's job to the detriment of one's health and well-being (Blair-Loy and Cech, 2017; Kossek et al., forthcoming).

Occupational cultures, such as academic culture, often socialize to believe success requires sacrificing their personal lives, which reinforces overworking (Blair-Loy and Cech, 2017; Kossek et al., 2001; NASEM, 2020). In addition, early-career scientists and physicians are often juggling romantic relationships, partnering, and starting a family, which are assumed to potentially harm future career prospects because they can distract from work roles (Kossek and Lee, 2020a). For example, a 2019 study showed that the rates of leaving the profession after the birth of a first child for academic STEMM women were double the rates for men (Oliveira et al., 2019).

Intersectionality and Work-life Research

Scholars have pointed to the growing relevance of diversity and inclusion concepts (Kossek and Lee, 2020c) and intersectionality theory to work-life research issues (Mor Barak, 2020),[3] and the concept of intersectionality is opening up new avenues for work-life research and policy. For example, the work-life issues of single Black women have been largely ignored by academic institutions that have often considered and prioritized work-life issues in terms of gender and overlooked race issues that intersect with gender (Creary, 2020).

It is important to examine intersectional work-life issues because underrepresented faculty, such as Women of Color, are more likely to report perceptions of work exclusion where they feel that their personal and professional needs and values are not being addressed (Mor Barak, 2020; Zimmerman et al., 2016). For example, national data show that a Black woman with a college degree in her midthirties to midforties is 15 percent less likely to be married than a white woman without a degree (Brookings Institution, 2017). This trend is exacerbated in less racially diverse rural and small city college towns where many academic institutions are located (Creary, 2020).

Work-life preferences for employer support intersect not only with race and gender but also with other forms of difference, such as parental status, disability, age, and career stage (Kossek and Lee, 2020a, 2020c). In spite of this, organizations and scholars have largely not attended to growing diversity and intersectionality in work-life needs (Kossek and Lee, 2020c; Mor Barak, 2020), which can impact how work-life boundaries are managed in racially and gender-imbalanced work units.

Boundary Management of Personal Identities in Gender and Racially Imbalanced Contexts

Whereas most research has focused on boundary management as a means to handle conflicting role demands, existing research also addresses the effect of boundaries on workplace relationships and employees' professional identities (Dumas et al., 2013; Dumas and Sanchez-Burks, 2015). Employees not only attend to whether the tasks associated with their work and family roles conflict

---

[3] The concept of intersectionality is a lens for understanding how social identities, especially for marginalized groups, relate to systems of authority and power.

but also whether aspects of their personal identities conflict with the accepted or desired norms for professionalism in their workplace. When women work in fields dominated by men, as many women in STEMM do, many report feeling that their gender is seen as incompatible with professional norms. As a result, their boundary management practices take the form of concealing aspects of their personal lives that highlight their gender or parental status if they are mothers (Cheryan et al., 2009; Jorgenson, 2002; Prokos and Padavic, 2002).

Similarly, work organizations often send the message to members of marginalized groups that they must alter their behavior to fit with professional norms (Ramarajan and Reid, 2020). As a result, Black, Latinx, and other People of Color are often intentional in managing the boundary between their personal and professional lives to preserve workplace relationships with dissimilar others (Dumas et al., 2013). For example, Black employees report refraining from disclosing personal information to their white coworkers because of concerns over career repercussions (Phillips et al., 2018). When they do disclose personal information, they may be careful to share only what will enhance their status at work and downplay their racial or gender category (Phillips et al., 2009; Yoshino, 2001). Research also indicates that refraining from discussing personal information at work, or strategically downplaying one's demographic categories, is also within the realm of managing the work-nonwork boundary.

Work-Life Policies and Practices Traditionally Offered by Academic Institutions

While most academic institutions believe they provide an environment that supports a healthy work-life balance, research suggests they generally fail to some degree (Kossek and Lee, 2020a, 2020c; Matthews, 2020). A few innovative programs have emerged that provide workload assistance to relieve time pressures, such as for physician scientists (Jagsi et al., 2018; Jones et al., 2019). Such programs may help to reduce the stigma of disclosing the caregiving responsibilities described above, as participants share experiences and increase awareness of the frequency and legitimacy of such caregiving challenges (Jones et al., 2020). However, with the exception of some of these newer piloted programs, which have yet to be fully integrated into academic institutions, far less attention has been devoted to using work-life policies to support the development of healthy work-life boundaries and cultures of well-being as a vehicle for faculty retention (Kossek and Lee, 2020b, 2020c).

The most common ways that academic institutions have responded to faculty work-life needs are (1) offering dual-career hiring to attract and retain academic faculty, with less consistent support for hiring nonacademic spouses; (2) offering childcare centers on campus, though spaces are often limited with long waiting lists, particularly for infant care; (3) allowing faculty to extend the tenure clock with parental leave; and (4) offering help with realtors and school information for faculty with children when hired (CUWFA, n.d.; Kossek and Lee, 2020a, 2020b, 2020c; Matthews, 2020; Schiebinger et al., 2008). While it seems less common for academic institutions to provide work flexibility for employees who are parents and those with eldercare demands, such as control over the timing of early morning or night classes and meetings, more evidence is needed to corroborate this view. An ongoing National Science Foundation study is exploring this observation to see if it does indeed hold true (Kossek et al., 2019–2021).

Supervisors and Peer Cultural Support Matters

A large body of work on supervisor support for family and personal life suggests it is likely that much of an academic department's support for how it accommodates family and personal life scheduling needs is often determined on an ad hoc decision-making basis by the department chair, resulting in wide variability (Kossek, 2005, 2006a, 2006b). However, evidence shows the benefits of strong, consistent leadership and an organizational culture that supports work-life issues. Meta-analyses show that when individuals perceive their supervisors as supporting work and family or personal roles, they are more likely to experience less work-family conflict and perceive their organizations as work-life supportive (Kossek et al., 2011).

Regarding eldercare and sandwiched care supports, given these are often outsourced to employee assistance firms, with universities often taking a hands-off approach, it is likely this support is also uneven in effectiveness, though once again this needs to be systematically investigated (CUWFA, n.d.). Support for the tensions of juggling dual academic careers that may vary in job security or career progress or for single parents is also limited (Thompson, 2020).

Systematic Work: Redesigned and Reduced-Load Work Options Overlooked

While academic institutions have overlooked adopting work redesign and cultural interventions to increase organizational support for work-life issues as a form of support for diversity and inclusion (Kossek, 2020), they do often provide leaves of absence for common work-life needs such as unexpected family care needs due to illness. It may be easier to offer faculty time off as a short-term solution rather than experiment with redesigning occupational work cultures and reducing job demands. Many private-sector employers offer customized, reduced-load work options to enable high-talent employees to experience a more balanced life during career advancement as a means of fostering sustainable careers and retaining employees (Kossek and Ollier-Malaterre, 2019). Faculty, however, are largely expected to self-manage and know how to create their own healthy boundaries. Overall, many academic institutions have yet to move work-life issues from the margins to the mainstream of job design and talent management strategies (Kossek et al., 2010).

Stepping out of the workforce for even a few years can risk career derailment and significantly decrease lifetime earnings with accrued pension effects from career gaps. Such trends have led scholars to depict women's careers as having "the sagging middles"; the tendency of many women to decrease hours and work productivity or leave the labor force after a first or second child (Goldin and Mitchell, 2017). Indeed, a study of pay equity of faculty from 1980 to 2004 found that gender pay gaps can be attributed to career interruptions and declines in accumulated human capital due to stepping out of the workforce or cutting back for children (Porter et al., 2008). These effects vary within STEMM disciplines.

POST-COVID-19 PANDEMIC LITERATURE: CHANGES TO BOUNDARIES, BOUNDARY CONTROL, AND WELL-BEING

Given the lead-time for publishing academic articles, few published studies directly examine work-life challenges arising from the COVID-19 pandemic for women faculty in STEMM. However, the common themes in the articles published during 2020 were consistent with findings in foundational work-family literature and, while not STEMM specific, with the literature on the work-life challenges of academic motherhood (Ward and Wolf-Wendel, 2012).

Rise in Childcare and Homeschool Demands and Increased Partner Tensions

As workplaces, schools, and childcare centers closed in response to the COVID-19 pandemic, many parents faced new and unusual dependent care and domestic demands, including homeschooling their children. With children and working parents in the home all day, parents had to reorganize caregiving time and work time. Several studies during spring and summer 2020 showed that caregiving time fell largely to mothers (Carlson et al., 2020a; Craig and Churchill, 2020; Myers et al., 2020; Shockley et al., 2020).

In one COVID-19–specific study, researchers analyzed data from the International Society for Stem Cell Research member survey (Kent et al., 2020), nearly 56 percent of which came from academics.[4] More than 85 percent of survey respondents reported increased caregiving responsibilities, and almost 50 percent of these respondents indicated that the additional family responsibilities disrupted their work. This trend was even greater among early-career faculty members, as 71 percent reported that their increased childcare responsibilities were hindering their work. The only reported home intervention for securing stretches of time to complete work was to trade off working shifts with a partner.

---

[4] Information about gender was not included in this survey (Kent et al., 2020).

Several studies have shown there are health and well-being implications of these unequal childcare responsibilities. One recent study found that for couples in which the wife was working remotely and taking on all of the childcare responsibilities, women reported the lowest family cohesion, highest relationship tension, and lowest job performance (Shockley et al., 2020). Similarly, a study of the rates of anxiety among physician mothers showed that 41 percent scored over the threshold for moderate or severe anxiety (Linos et al., 2020). Other studies have also found that the mental health of working mothers has suffered during the pandemic (Zamarro and Prado, 2020). In addition, women business leaders and women employees reported greater work-related stress compared with their counterparts who are men (Catalyst, 2020a) (see Chapter 7 for more on mental health and well-being).

To cope with additional caregiving demands, women are reducing their work hours (Madgavkar et al., 2020). For women in STEMM with children or other dependent-care responsibilities, many had significantly less time in the day to network and engage in collaborations because of increased nonwork tasks (Heggeness, 2020; Kossek and Lee, 2020b; Myers et al., 2020). One study of dual-earner, married couples with children found that for parents in telecommuting-capable jobs with children between 1 and 5 years old, mothers report nearly 4.5 times larger reductions in work hours than fathers (Collins et al., 2020). Moreover, in another study of 25 U.S. women faculty members and 55 Italian women faculty members, all of whom had children, the women reported a perceived cognitive deficit from managing the demands of children all day (Minello, 2020). These responses were consistent with research showing that blurred work-nonwork boundaries are associated with increased work-family conflict (Hecht and Allen, 2009; Kossek et al., 2006; Krukowski et al., 2020).

A potential positive change was that one-third of employees who are men and nearly two-thirds of business leaders who are men reported taking on a more equitable share of chores at home in a survey (Catalyst, 2020a). Emerging work from several nations suggest that men have started shouldering more caregiving and child-rearing duties, a view corroborated by their partners (Carlson et al., 2020b; Savage, 2020; van Veen and Wijnants, 2020). A study by a consumer marketing group found that 62 percent of men wanted to keep working at home specifically because it increased family time (Fluent, Inc., 2020). In addition, slightly more women than men expressed the belief that the new working environment during the COVID-19 pandemic may provide more flexibility in work-life balance and control over their schedules in the future (Catalyst, 2020a).

Work-Life Effects of the COVID-19 Pandemic on Faculty of Color

While the committee could not find refereed empirical scholarly literature on how the COVID-19 pandemic has affected the work-life challenges specific to STEMM women Faculty of Color, there were media reports of disparate negative health, career, and work-life effects. Many news reports provided anecdotal evidence that the pandemic negatively affected the well-being of many Faculty of Color compared with their white counterparts. Faculty of Color were more likely to have or know a family member or friend who got ill or died from the virus than white faculty. The COVID-19 pandemic also made it difficult for more junior faculty hires to find housing, which became more expensive and more difficult to secure (Brooks, 2020). The tightening labor market, rescinded new-hire positions, institutional layoffs, and dissolution or reorganization of departments to manage the decline in student enrollments negatively affected the careers of Faculty of Color (Aviles, 2020) (see Chapter 3 for more on academic careers).

CONCLUSIONS

Women faculty in STEMM faced unique challenges resulting from the COVID-19 pandemic related to juggling growing second-shift challenges juxtaposed with increased boundary permeability, rising workloads, and persistent ideal-worker cultures. Remote work can be a double-edged sword for women's careers (Kossek et al., 2014). For example, while it can facilitate the management of work-family roles, it also increases multitasking, process losses from switching frequently between tasks, and interruptions and extended work availability that may harm mental health and well-being. Data from the survey discussed in Chapter 2 identified strategies that women faculty use to manage boundaries during the COVID-19 pandemic. A potential positive outcome from the enforced work-from-home arrangements associated with the COVID-19 pandemic may be that individuals developed new skills in setting technological, temporal, and spatial adjustments to manage boundaries. These new skills in boundary management may continue to be useful to their career development in the long term. Adjustments resulting from the COVID-19 pandemic present the opportunity to compare the benefits and detriments of different boundary management styles.

Finally, there are likely differences at many levels between individuals whose careers were derailed and those who are more successful. In other words, the evidence was not yet established to understand how differences in departmental supervisor, discipline, and university context influenced the experiences of women in STEMM around overwork culture and boundary norms during COVID-19. Moreover, how intersectionality influences the experiences of Women of Color in STEMM remains undetermined at this time.

5

Collaboration, Networks, and Role of Professional Organizations[1]

INTRODUCTION

The COVID-19 pandemic had a substantial reach into many aspects of academic science, technology, engineering, mathematics, and medicine (STEMM) life in 2020, and research collaborations, mentoring and sponsoring relationships, networks, and professional organizations were not spared (Al-Omoush et al., 2020; Kramer, 2020b). The effects on the workforce were documented in published studies, and STEMM experts continue to highlight the differentially gendered effects on science and scientific collaborations (Buckee et al., 2020; Gruber et al., 2020; Kramer, 2020b; Minello, 2020; Myers et al., 2020) and in service activities such as mentoring and advising at work (Kramer, 2020b). While there may be benefits of video conferencing and technology that provide a means to maintain collaboration and communication among scientists during a pandemic (Korbel and Stegle, 2020), evidence also suggests that shifting priorities both within households and at work for academic women during the COVID-19 pandemic has restricted their ability to engage in collaborative work and networking at the same level they might have been engaged prepandemic (UW System, 2020; Zimmer, 2020) (see Chapter 4 for more on work-life balance). In light of this challenge, institutions are being encouraged to understand the explicit and implicit mechanisms of the COVID-19 pandemic on shifting norms in collaboration and networking for women STEMM academics. If no action is taken, the differential effect that the pandemic has had specifically on academic women's

---

[1] This chapter is primarily based on the commissioned paper "The Impact of COVID-19 on Collaboration, Mentorship and Sponsorship, and Role of Networks and Professional Organizations," by Misty Heggeness and Rochelle Williams.

ability to collaborate, mentor, and network will continue going forward (Kramer, 2020b), potentially worsening already existing gender-based inequalities.

This chapter focuses on how the COVID-19 pandemic has affected collaboration and networking, as well as the role professional organizations have served, for academic women during 2020. A description of how the authors of the commissioned paper on which this chapter is based selected the materials is available in Appendix C.

HISTORICAL EVENTS AND THE IMPACTS ON COLLABORATIONS AND NETWORKING

The COVID-19 pandemic is not the first time a major global crisis has shifted norms associated with academic collaboration and networking at universities. After World War II, international collaborations on university campuses and interactions among academic colleagues globally were stunted, resulting in reduced productivity, as measured by publications and patent awards, and a slowing of new and novel advancements along the scientific research frontier (Iaria et al., 2018). The aftermath of September 11, 2001, drove changes to how colleges and universities accept students, postdoctoral students, and faculty from abroad. International student migration dropped post-9/11 and shifted worldwide student mobility trends (Johnson, 2018). Barriers to visa and work permits were reported to stifle international collaboration and engagements as well as advancements in innovation and productivity (Chellaraj et al., 2005). Four years later, Hurricane Katrina forced students and faculty to relocate to other campuses, which influenced both the workflow and psyche of faculty and students, as well as of those who collaborated and worked with them (AAUP, 2007).

These historic experiences indicate that methods of collaborating among scientists are hindered and stunted by crisis events, through either human- and nature-made restrictions or policies restricting communication and engagement standards with potential collaborators. These barriers to collaborative networks have long-term effects that bleed into future success, including achieving tenure and reducing potential future collaborations and research output of academics and STEMM researchers (Chai and Freeman, 2019; Chellaraj et al., 2005; Iaria et al., 2018). Restricting or canceling participation in professional organization conferences, which occurred regularly in early spring 2020, has also been shown to alter future collaborations negatively (Chai and Freeman, 2019). Nonetheless, professional organizations can play a vital role in the interinstitutional efforts required to meet the needs of their membership and the society at large.

EFFECTS OF THE COVID-19 PANDEMIC ON COLLABORATIONS AND NETWORKING

With advances in technology and cloud computing, collaborations can more easily flow across state and country lines and, at first glance, it may appear that these technologies have mitigated or reduced the damage to collaborative work during the COVID-19 pandemic (Apuzzo and Kirkpatrick, 2020). However, collaborators have nonetheless had to adjust to the sudden elimination of in-person engagement as a result of public health policies enacted during the COVID-19 pandemic (CDC, 2020b, 2020c).

There are general approaches to collaborative research that changed in response to losing the ability to physically meet in person. One preexisting approach was to use online platforms such as Zoom, Webex, or Microsoft Teams meetings; emails; cloud-computing shared spaces; and other digital formats (Clark, 2020). By taking this approach, a relatively small decline in collaborative research might have occurred. However, family obligations during the COVID-19 pandemic may deter or hinder the ability to collaborate even if remote options are open and accessible (Myers et al., 2020). (See Chapter 4 for more on how family obligations are affecting women STEMM faculty.)

Some collaborative research projects, however, cannot be conducted over an internet connection and instead require face-to-face interaction to thrive and survive. Such projects include those requiring fieldwork or experimental or bench/wet-lab research. Indeed, collaborations requiring expensive and exclusive laboratory equipment to advance cannot easily be replicated in a home office. In these situations, collaborations have been slowed or put on hold (Radecki and Schonfeld, 2020). These delays in timing have the potential to sour time-sensitive laboratory projects and put a strain on demonstrating outcomes of grants and other types of research funding. In addition, as laboratories adapt to COVID-19 pandemic safety measures, they have to adjust to doing more with fewer interpersonal interactions because of social distancing and other measures (Brockmeier, 2020; Radecki and Schonfeld, 2020; Schiffer and Walsch, 2020; Schmidt, 2020).

Even though federal funding agencies have been providing support to scientists during the COVID-19 pandemic,[2] the effect of the COVID-19 pandemic on collaborative networks does have the potential to negatively affect future grants for projects that may have been funded based on current collaborative research that was slowed or terminated (Heidt, 2020; Whitlock, 2020; Yeager, 2020).[3] These challenges to data collection and essential travel have put research at risk and reduced or slowed the speed of research and publication of results (see Box 5-1). This, in turn, has the potential to negatively affect tenure and promotion (see Chapter 3 for more on this subject).

---

[2] For more information, see https://grants.nih.gov/policy/natural-disasters/corona-virus.htm.

[3] For example, the U.S. Census Bureau had to delay its door-to-door data collection program for the 2020 Census by several months because the pandemic made it risky for enumerators to collect responses in person (U.S. Census Bureau, 2020).

> **BOX 5-1**
> **Effects of the COVID-19 Pandemic on**
> **International and Distanced Collaborations**
>
> International and distanced collaborations involve additional hurdles for keeping teams on track and STEMM fields advancing in normal times. With the arrival of COVID-19, many collaborative endeavors were unprepared for a disruption such as the pandemic and "the pandemic ... robbed many laboratories of international researchers and the diverse skills and viewpoints they bring" (Woolston, 2020b). Field experiments, which by nature are conducted on-site within local communities, have been lost to the extent travel of principal investigators and other staff conducting the experiments was essential to their advancement (Devi, 2020). *Science* reported that "the coronavirus pandemic had already canceled one summer field research season. Now it [had] come for another: the Antarctic summer ... the United States and United Kingdom would put most of their planned Antarctic research into deep freeze, including their ambitious joint campaign to study Thwaites Glacier, the Antarctic ice sheet most at risk of near-term melting" (Voosen, 2020).

Effects of COVID-19 on Collaborations and Networking for Women Faculty

During the course of 2020, it became beneficial for institutions to anticipate areas where women's professional productivity and advancement could be negatively affected by the challenges of conducting collaborative research (Andersen et al., 2020; Muric et al., 2020). When considering specific disciplines, like Earth and space scientists represented through the American Geophysical Union and life scientists, virtual collaborations have proved incredibly beneficial for journal clubs, meeting collaborators, workshops, and conferences, among other collaborative mechanisms (Korbel and Stegle, 2020; Wooden and Hanson, 2020). However, preliminary evidence from 2020 suggests that the COVID-19 pandemic affected women's ability to engage actively in collaborations. Studies suggest, too, that team size has decreased during 2020 and that women's shares of first authorships, last authorships, and general representation per author group have decreased during the COVID-19 pandemic (Andersen et al., 2020; Fry et al., 2020) (see Chapter 3 for more on publication).

Buckee et al. (2020) describe the complex situation of women scientists studying the COVID-19 pandemic, particularly Women of Color, who are often not noticed by the media and yet play a major role in getting the work done and advancing breakthrough science. To address these issues, Black engineering faculty from multiple institutions formed the grassroots organization Black in Engineering to draft and disseminate the call-to-action report *On Becoming an Anti-Racist University* (Black in Engineering, 2020). Drafting this call to action took considerable time away from teaching preparation, mentoring, and research in which these scholars would have otherwise been engaged.

Additional Effects for Women Faculty with Children and Caregiving Responsibilities

Many parents or caregivers are unable to participate in international or distanced collaborations, even virtually, because of the time commitment required that competes with the need to serve as a homeschool teacher to school-age children in their care, as discussed in Chapters 2 and 4. This issue is not new to 2020. In March 2018, the Working Group of Mothers in Science published an opinion piece in the *Proceedings of the National Academy of Sciences of the United States of America* titled "How to Tackle the Childcare–Conference Conundrum." The group compiled four recommendations directed toward facilitators of collaborations, such as research societies and conference organizers, titled CARE, for Childcare, Accommodate families, Resources, Establish social networks. These recommendations were made as a direct response to the parent-researcher struggling to attend key conferences and further their careers while securing care for children. The CARE recommendations are intended to enable women in academia to have equitable opportunities to make contact with representatives from funding agencies, communicate new research and discoveries, form collaborations, and attract new members to research teams (Calisi et al., 2018). These considerations are even more relevant during the COVID-19 pandemic, when mothers' workload within their households has increased substantially, as discussed in Chapter 4 (Del Boca et al., 2020; Sevilla and Smith, 2020).

EFFECTS OF THE COVID-19 PANDEMIC ON PROFESSIONAL ORGANIZATIONS AND NETWORKS

Professional societies and networks can serve as conduits to standing up programs and policies that help mitigate the loss of collaborative networking among women scientists, as these organizations have historically responded to national crises by equipping their members to meet societal needs (Morris and Washington, 2018). For example, in 2005, the Association of American Medical Colleges (AAMC) organized an emergency conference call with all U.S. medical school deans to coordinate the response from the National Institutes of Health and academic medicine to meet the health-care needs of patients affected by Hurricane Katrina. Additionally, AAMC created a website to coordinate offers of housing and laboratory space for researchers and students displaced by the storm (Cohen, 2005). Professional organizations also advocated for the needs of the scientific workforce and have continued to engage in advocacy and outreach efforts on behalf of the scientific community and related groups (Segarra et al., 2020).

Professional organizations can also play a role in resource sharing and networking components that may be pivotal to career advancement and ongoing education. Even with recent technological advances that allow for remote communication, physical attendance at various venues, including conferences, lectures, and networking events, are primary means by which scholars build their research programs (Segarra et al., 2020). As women continue to face inequitable obstacles to fully attending and participating in networking and development experiences because of responsibilities related to children and other family obligations (e.g., eldercare), professional organizations now have an opportunity to reimagine the structure of membership, conferences, and events. Fortunately, technology now provides alternate ways to do this, and by using an inclusive lens, organizations can broaden access by reimagining conferences and meetings.

During the social distancing guidelines and stay-at-home orders in place in 2020, professional organizations catered to an emerging set of member needs around collaboration, mentorship, and sponsorship while navigating constraints arising from the cancellation of in-person events, decreased philanthropic support, and less income typically supplied by membership fees. The financial losses that many of these organizations have suffered hindered the ability of professional trade and professional associations to establish alternative approaches to making their activities more widely available during the COVID-19 pandemic. These organizations remain vulnerable to severe financial issues as the COVID-19 pandemic continues to limit member participation. Despite financial constraints, professional organizations and networks in STEMM are being called to remain resolute in their tenets to promote professional excellence and help promote change at the intersection of a global health crisis and a fight for racial justice within the United States (Community Brands, 2020; SWE, 2020). Professional organizations responded quickly to member and societal needs, while they strived not to undo years of effort in creating diverse and equitable opportunities for persons with identities traditionally excluded from STEMM.

Online communities serve as one mechanism in which professional organizations can promote inclusive online environments that drive conversation around gender equity. Higher Logic, an engagement platform that delivers online communities and communications software to more than 3,000 customers, primarily community, advocacy, and professional associations, reported that online community engagement increased between February and April 2020 (Bell, 2020). For example, the National Society of Professional Engineers (NSPE) and American Society for Microbiology (ASM) opened their online discussion forums to the public to facilitate communication and engagement with leading NSPE and ASM members, respectively (ASM, 2020; NSPE, 2020). While not explicitly tied to gender equity efforts, members at the American Occupational Therapy Association (AOTA) are using their online community as a source of connection and education. In all, overall logins at AOTA increased 109 percent between mid-March 2020 and mid-April 2020. They also found that new member logins, those utilizing the community for the first time, increased by 149 percent. While in the online community, online discussion posts increased 48 percent, views of library content saw a 100 percent increase, and library downloads increased 122 percent between February 2020 and April 2020 (Bell, 2020).

Given that professional organizations often serve as an intellectual resource for STEMM communities, this suggests that members are relying on their association communities as vital sources of connection and information. The broader access to membership benefits such as online discussions and resource libraries could prove to be beneficial to the community at large.

Leveraging Working Groups and STEMM Networks to Promote Gender Equity during the COVID-19 Pandemic

One way in which professional organizations can better highlight their efforts toward supporting women in academic STEMM during the COVID-19 pandemic is via working groups. Working groups, also called affinity groups or divisions, are organization-recognized microcommunities that can serve to promote diversity and inclusion efforts and allow for networking, mentoring relationships, and other opportunities for professional and personal development (Taylor, 2019). Historically, affinity groups were centered on race or gender, but these groups are increasingly being created for those sharing other characteristics, such as age, sexual orientation, and disability status. Groups for women in professional organizations typically address gender equity, recruitment and retention, awards and recognition, and career advancement (AAMC, 2020; Taylor, 2019).

When considering the role of gender-centered affinity groups or working groups housed at professional organizations, evidence on the overarching roles of these groups suggests that these groups have been deployed during the COVID-19 pandemic to advocate for the women members within the organization, support the inclusion and raise the visibility of women in virtual meetings spaces, and ensure organizations take intersectionality into account as they develop interventions to support women's advancement in the field.

To further improve efforts to support academic women in STEMM during and after the COVID-19 pandemic, evidence suggests that creating partnerships with networks that specialize in addressing gender equity issues in STEMM could further amplify gender-equity work at professional organizations (ARC, 2020; Aspire, 2020; NIH, 2020a). STEMM networks have been leading discussions on ensuring equity for women in academia during the COVID-19 pandemic, but their reach is often limited to those who already know of the network. Intentional partnerships between STEMM networks and professional organizations can simultaneously heighten the visibility of women-centered working groups within professional organizations and broaden the reach of STEMM networks. Given that one high-impact function of many professional organizations is sponsorship (e.g., nomination for awards, leadership opportunities, scientific plenaries, reviews, visiting professorships) (Cree-Green et al., 2020), coupled with technical and scientific work relying on research teams, groups, and the cooperation of people (Fox, 2001), these bodies have an opportunity to shift how mentoring and sponsorship is viewed both within and outside of academic institutions for women in STEMM.

To develop larger-scale interinstitutional change, there are opportunities for institutions and professional organizations to engage a number of equity-centered networks in STEMM. For example, over the past 20 years the National Science Foundation (NSF) Increasing the Participation and Advancement of Women in Academic Science and Engineering Careers (ADVANCE) program has provided funding to support the implementation of evidence-based systemic change strategies that specifically promote equity for women STEM faculty in academic workplaces and the academic profession.[4] In 2010, the ADVANCE Implementation Mentors (AIM) Network was formed to establish a common mentoring network for ADVANCE program coordinators and project directors at all developmental stages of ADVANCE grants with the purpose of answering questions and providing support, sharing promising practices, and establishing a common resource base. With membership of more than 80 program directors and managers, AIM is a community of practice that accelerates and disseminates the work of NSF ADVANCE. More recently, NSF established the ADVANCE Research and Coordination Network in 2017 to facilitate authentic, intentional dialogue between researchers and practitioners, connect inclusiveness to organizational principles and practices, and account for and incorporate intersectional perspectives throughout the network members' work. Both networks have been instrumental in convening national audiences to discuss the issues women faculty in STEMM are facing and the resources they need to survive the COVID-19 pandemic.

EFFECTS OF THE COVID-19 PANDEMIC ON CONFERENCING

Professional societies and academic institutions, through the conferences they hold, can provide women STEMM professionals with the opportunity for development of national recognition and academic relationships beyond their home institution (Cree-Green et al., 2020). However, even prior to the COVID-19 pandemic, researchers had begun calling for a paradigm shift in how scientific conferences are conducted to reduce their contribution to climate change by reducing travel-associated emissions (Levine et al., 2019). While professional conferences were canceled in the first half of March 2020 (Benchekroun and Kuepper, 2020), they quickly started converting to virtual programs starting in the latter half of March (ASAE, 2020), and virtual conferencing became the new normal in 2020. In the wake of a global pandemic, in-person seminars transitioned to virtual seminars and in-person coffee breaks, and happy hours converted into virtual meetups on video-conferencing platforms, allowing for a broader participation from international colleagues. Those in privileged situations—those without care responsibilities and with reliable access to high-speed

---

[4] Additional information about the NSF ADVANCE program is available at https://www.nsf.gov/crssprgm/advance/index.jsp.

internet connections—navigated new paths forward despite the need for social distancing and the inability to meet physically in a central location. Universities and professional organizations adjusted budgets to accommodate the financial requirements needed for this virtual transition and lower enrollments (Hemelt and Stange, 2020).

In many ways, the move to virtual platforms provided new opportunities for those with limited travel funds to participate in conferences and seminars in which they otherwise would not have been able to take part, particularly for early-career individuals (Adams, 2020; Segarra et al., 2020). Additionally, virtual seminars gave institutions the ability to invite speakers to present their research they would not have otherwise been able to invite because of limited travel budgets, and individuals across the country gained opportunities to participate in some institution-specific seminars that they would otherwise have not attended (Kalia et al., 2020; Segarra et al., 2020). However, while these expansions opened access, they did not eliminate the challenges of attending these events while engaged in childcare activities within their households, which many women in academic STEMM have experienced during the COVID-19 pandemic. Moreover, the conference experience is different because networking is limited by one's ability to engage with other scientists via a virtual platform.

Building Virtual Environments Conducive to Collaboration and Networking

The cancellation of in-person conferences may have led an entire cohort of early-career scientists and academics to lose out on networking opportunities if not for the development and implementation of new technologies by conferences during 2020 (Benchekroun and Kuepper, 2020). The longer-term effects of the changes in conferences during 2020 on the careers of early-career scientists going forward are yet to be realized. Institutions have the opportunity now to work through how this may be affecting their own early-career academics and plan alternative options to help mitigate the negative effects.

Prior to the COVID-19 pandemic, researchers investigating the shortcomings of scientific conferences found it imperative that organizers of virtual meetings and events improve strategies for facilitating digital connections (Avery-Gomm et al., 2016; Sarabipour et al., 2020). Specifically, they found that scientific organizations were utilizing features such as Slack to facilitate both group and one-on-one discussions during and after meetings, along with incorporating Twitter to deliver poster sessions (Avery-Gomm et al., 2016; RSC, 2019; Sarabipour et al., 2020). However, when professional organizations shifted to virtual events as a result of the COVID-19 pandemic, they were cautioned to keep membership engagement in a central location to ensure members can readily find information and directly network with the organization and other members (Community Brands, 2020). As such, efforts to transition conferences and events to inclusive virtual spaces hinged on investing in immersive and interactive experiences that promote collaboration and networking (Sarabipour et al., 2020). At the end of 2020, the question still remained as to how virtual meeting components (e.g., ability for participants to engage one-on-one or in a group using video instead of chat, informal meetups, and social interaction) directly impact women's participation, advancement, safety, and ability to collaborate and network with their peers in this new setting.[5]

Bias in Virtual Environments

In July 2020, the Society of Women Engineers (SWE) published a survey report titled *Impact of COVID-19 on Women in Engineering and Technology*. The survey of its members, open between June 3, 2020, and June 15, 2020, examined how the COVID-19 pandemic affected their personal and professional lives. Analysis focused on responses received from women and genderqueer/nonbinary people who made up 98 percent of the total respondents. More than a third of the respondents identified as working professionals, a quarter of the respondents were People of Color, and nearly a third of the SWE professionals who responded reported experiencing bias during virtual meetings in the form of getting talked over, interrupted, or ignored more frequently during virtual meetings than those held in person (SWE, 2020). When disaggregating the data by age group, a higher proportion of younger SWE professionals (aged 18 to 24 and 25 to 34 years) than SWE professionals aged 55 to 64 years reported getting ignored (35 percent), interrupted (38 percent), and talked over (22 percent) more frequently during online meetings than those held in person.

In a report published by Catalyst in June 2020, 45 percent of women business leaders reported that it was difficult for women to speak up in virtual meetings—42 percent of men business leaders agreed with this observation—and one in five women had recently felt ignored and overlooked by coworkers during video calls (Catalyst, 2020a). SWE Professionals of Color were as likely to report similar frequencies of getting interrupted, talked over, and ignored in virtual meetings as their white peers, while Women of Color and genderqueer/nonbinary SWE Professionals of Color disproportionately reported other concerns, such as losing their job as a result of the economic impact of the COVID-19 pandemic on their employer.

A review of the public-facing web pages of 246 STEMM professional organizations did not show evidence of the steps organizations are taking to combat the bias women experience in virtual meetings and events or how they were being intentional about highlighting presentations, keynotes, and resources written by women during the COVID-19 pandemic.[6] In a study conducted to review the

---

[5] There are examples of large-scale, virtual professional society meetings, such as the American Geophysical Union 2020 meeting held during December 2020 that used several synchronous and asynchronous modes of engagement.

[6] More information about this review can be found in Appendix C.

features of scientific conferences, researchers found that 97 percent of 270 scientific conferences they examined lacked a statement of gender balance or diversity (Sarabipour et al., 2020). Additionally, they found that out of the meetings reporting names of chairs, organizers, and invited speakers online, only 43 percent and 34 percent of conferences achieved gender parity for conference chairs and session chairs, respectively; 41 percent achieved gender parity for conference organizers or steering committees; 32 percent and 34 percent achieved gender parity for keynote and plenary speakers, respectively; and only 17 percent had equal numbers of men and women as invited or featured speakers (Sarabipour et al., 2020).

Researchers reviewing medical conferences also reported that women are underrepresented among conference and symposium session chairs, plenary or keynote speakers, invited lecturers, or as panelists in a broad range of academic meetings (Gerull et al., 2019; Larson et al., 2019; Ruzycki et al., 2019). Similarly, the American Geophysical Union investigated the chances of scientists from groups that are underrepresented in Earth and space sciences being given speaking opportunities, compared with other applicants, at their annual fall meetings from 2014 to 2017. The results indicated that first authors from Communities of Color contributed 7.7 percent of all the abstracts in the sample (n = 2,981) and that these applicants were disproportionately students or early-career scientists and were less likely to be invited to give presentations (Ford et al., 2019).

THE ROLE OF MENTORSHIP[7] AND SPONSORSHIP[8] DURING THE COVID-19 PANDEMIC

The advancement gap between men and women in academic STEMM developed because the systems have inherent performance support and reward bias built into them that may require additional guidance and navigation for women (Castilla and Benard, 2010; Roper, 2019). When women have sponsors, it can narrow the advancement gap between women and men in academic STEMM (NASEM, 2020; Patton et al., 2017).[9] Research has shown that mentors can positively affect the career outcomes and advancement of academic women (Ginther et al., 2020), but it also suggests that mentoring and sponsoring relationships serve as a source of men's invisible advantage in STEMM, given women's lack of access to senior academics who would serve as sponsors (O'Connor et al., 2020). Studies have shown that women benefit from multiple mentors of all genders and

---

[7] *Mentorship* is a professional working alliance in which individuals work together over time to support the personal and professional growth, development, and success of the relational partnerships through the provision of career and psychosocial support (NASEM, 2019c).

[8] *Sponsorship* is a potential career support function that involves a senior person publicly acknowledging the achievements of and advocating for a mentee or protégé (NASEM, 2019c).

[9] A *sponsor* is typically a senior-level person who advocates for their protégés.

that mentees receive a different experience from mentors who identify as men versus those who identify as women (O'Brien et al., 2010).

The importance of having multiple mentors for women STEMM academics is essential during difficult times, which is why researchers have suggested institutions proactively support, encourage, and develop mentorship and sponsorship programs for their women faculty and staff during and after the COVID-19 pandemic (Mickey et al., 2020). Furthermore, peer support network interventions may help address reduced opportunities for interactions between colleagues, increased social isolation, and reduced mentoring opportunities related to working from home. These factors affect both the psychological well-being and career outcomes of women faculty, as discussed further in Chapter 7. For example, work is now being conducted with social media peer groups for physician mothers (Yank et al., 2019).

There was limited information available about changes in mentorship during 2020. In a survey circulated among life scientists in eight countries, including the United States, from April 15 to 23, 2020, almost 48 percent of respondents said that communication with their supervisor, mentor, or manager had remained consistent, while 22 percent of respondents said that their communication had increased, indicating the value and benefit of video conferencing and current technology (Korbel and Stegle, 2020).

While it is important for women faculty to continue to have access to mentors and sponsors during the COVID-19 pandemic, it is equally important to interrogate the systems that cause women to be overmentored and undersponsored (Ibarra et al., 2010). When considering how gendered faculty networks have affected the retention of women faculty in STEMM, the role of mentorship and sponsorship during COVID-19 appears to be two-fold: to support women faculty in navigating the systemic, professional, and personal challenges the evidence suggests they will encounter as a result of the COVID-19 pandemic, and to provide guidance and advocacy as they expand their networks.

CONCLUSION

The confluence of events that took place in 2020 highlighted the importance of professional organizations and networks using intentional, intersectional, and inclusive lenses to ascertain the range of opportunities and approaches available to build STEMM capacity, diversify STEMM fields, and meet the needs of academic women. Federally funded endeavors such as the NSF's ADVANCE program assisted professional organizations in magnifying policies and practices that not only support equity and inclusion but also mitigate the systemic factors that create inequities in the academic profession for women (NSF, 2020). To meet the needs of members navigating the COVID-19 pandemic in 2020, many professional societies and STEMM networks transitioned their in-person events to virtual experiences, adjusted various submission deadlines to accommodate for continued uncertainty, and advocated both locally and nationally on behalf of their constituents (ASAE, 2020; Community Brands, 2020). Because of the developing and ongoing nature of the COVID-19 pandemic, there will be an opportunity for future, systematic studies of how STEMM networks and professional organizations responded and functioned during the COVID-19 pandemic, particularly regarding who they serve, what they do, and with whom they collaborate.

6

Academic Leadership and Decision-Making[1]

INTRODUCTION

This chapter explores academic leadership and decision-making during the COVID-19 pandemic in 2020. As was documented in the National Academies of Sciences, Engineering, and Medicine report *Promising Practices for Addressing the Underrepresentation of Women in Science, Engineering, and Medicine: Opening Doors*, women have a long history of underrepresentation in academic institutional leadership roles (Glazer-Raymo, 2001; NASEM, 2020), a situation that has contributed to decisions that favor men and has created organizational structures that appear gender neutral but that are biased to favor men (Bilimoria and Liang, 2012). Various studies have explored and documented the reasons for women's historical underrepresentation in academic leadership positions. These reasons include the "chilly climate," where women are treated in a discriminatory fashion that becomes even "colder at the top"; embedded attitudes in academia favoring men's advancement (Allan, 2011; Dean et al., 2009; Eddy et al., 2017; Glazer-Raymo, 2001); and organizational work policies that make it challenging for women to succeed, such as tenure policies that make it challenging for women to have families. Universities have given relatively little attention to leadership development interventions to promote family and work-life supportive supervisory behaviors, which have been shown to be effective in randomized controlled trial experiments in other settings (Hammer et al., 2011; Kossek et al., 2019). In fact, data for women presidents and provosts demonstrate that they are less likely than men in these roles to be married or have children and more likely to have

---

[1] This chapter is primarily based on the commissioned paper "The Impact of COVID-19 on Academic Leadership and Decision-Making," by Adrianna Kezar.

altered their career for a dependent or spouse, as discussed in Chapter 4, suggesting that women find it more difficult to balance family and leadership roles (ACE, 2017). Women also have less access to networks that would help them to move up in the ranks of administration and are less likely to fit neatly into male cliques (Dean et al., 2009; Glazer-Raymo, 2001). In addition, societal stereotyping of men and women favors male traits in institutional leaders (Allan, 2011; Dean et al., 2009; Eddy et al., 2017; Glazer-Raymo, 2001).

Women of Color face additional challenges compared with white women. One study (Bridges et al., 2007) cited biased perceptions of Leaders of Color and their capacity to lead, which is often the result of conscious or unconscious reliance on existing stereotypes. Women of Color leaders in academia report tokenism and stereotyping as contributing to isolation, loneliness, and burnout (Bridges et al., 2007). Bias in the application of evaluation criteria and the tenure processes may account for inequities in Women of Color entering academic leadership. These are a sampling of the documented disparities that affect women faculty in particular but have outsized impacts later as they rise into leadership roles. These barriers, which were also discussed in Chapters 3, 4, and 5, are critical as they shape the pool for women leaders.

Women's representation in leadership is critical for closing equity gaps and making institutions more equitable workplaces (Bilen-Green and Froelich, 2010). Scholars have hypothesized that more women in strategic leadership positions would ameliorate work policy obstacles, given their experience of these barriers, and they would likely improve networking possibilities that might facilitate more equal participation of women within the academy (Bilen-Green et al., 2008; Langan, 2019). Data support this hypothesis, with investigators identifying relationships between the prevalence of women in strategic leadership positions and the associated impact on support for women in various professorial ranks. Research also documents that having a woman president resulted in more women faculty in full professor and tenure-track appointments (Bilen-Green et al., 2008).

Data on department chairs in economics, sociology, accounting, and political science from 200 institutions over 35 years show that woman department chairs narrowed three gender gaps (Langan, 2019): (1) assistant professors who work more years under a woman department chair have smaller gender gaps in publication and tenure; (2) the gender earnings gap decreases in the years after a woman replaces a man as a department chair; and (3) when a woman replaces a man as department chair, the number of women incoming graduate students increased by 10 percent without affecting the number of men. As women have moved into leadership roles, they have addressed work-life challenges tied to the male norms that dominate workplaces, including those that affect tenure clock provision and other work-life policies, as discussed in Chapter 4 (Wolfinger et al., 2008).

These are examples of a larger body of work suggesting that the representation of women in leadership has important and meaningful implications for creating equitable college and university environments and reversing gender-neutral policies that have had detrimental impacts on women. The collective data about women in higher education leadership roles suggest long-term underrepresentation has negatively affected progress toward equity, since data suggest women are more likely to make progress on these issues than men (Bilen-Green et al., 2008; Langan, 2019).

EFFECTS OF THE COVID-19 PANDEMIC ON WOMEN IN ACADEMIC LEADERSHIP POSITIONS

The COVID-19 pandemic appears to have had pronounced and immediate negative effects on women, as economic data appear to show (see Chapter 3). Various international policy organizations suggested that businesses, industries, governments, and other groups should be attentive and respond to these gender inequities (ILO, 2020).

The history of gender underrepresentation in leadership, gender inequities in academic decision-making, Gig Academy context,[2] the global recession, and the effects of the COVID-19 pandemic outside higher education suggest that the COVID-19 pandemic will likely exacerbate long-standing gender inequalities for women's advancement into leadership as well as decisions that shape gender inequalities. An exploration of institutional decision-making under academic capitalism and the Gig Academy is available in Box 6-1. Women's advancement into tenure-track faculty positions has been significantly altered by academic capitalism and the Gig Academy. These trends are also likely responsible for the slowdown in diversifying leadership such as provosts and board members where there has been little progress on gender parity. All these forces culminate in a set of reactions that played out on campuses during 2020.

COVID-19 PANDEMIC DECISION-MAKING AND EFFECTS ON GENDER INEQUALITIES

With the Gig Academy (see Box 6-1) pushing decision-making more to academic governing boards and administrations that are largely white and male, campuses had already seen a regression on gender equity that was further exacerbated during the COVID-19 pandemic (Flaherty, 2020b, 2020c, 2020f). Shared governance has been in decline for years but has now receded even more on most campuses. Since the start of the COVID-19 pandemic, there have been dozens of examples of reported overreach among governing boards making unilateral decisions without input from faculty, staff, and sometimes even the administration (Flaherty, 2020c, 2020f; Friga, 2020). Significant financial decisions have

---

[2] *Gig Academy* is a term coined by Kezar et al. (2019) to capture the ways that corporate gig economy practices, such as hiring contingent labor or outsourcing, have been adopted by colleges and universities across the country.

BOX 6-1
Changing Nature of Decision-Making Under Academic Capitalism and the Gig Academy

Part of the lack of progress of women moving into leadership during the past few decades can be associated with the rise of academic capitalism (starting in the mid-1980s) and the Gig Academy (starting around 2000) (Kezar et al., 2019; Slaughter and Rhoades, 2004). Academic capitalism prioritizes the marketization, individualism, and privatization of institutional operations as organizing principles for higher education. This organizing principle favors managers, market-based interests, hierarchy, and elite interests, while simultaneously moving away from a public or collective good, worker empowerment and self-organization, and worker participation in decision-making and the community. Marketization is promoted by adopting a corporate logic; that is, universities are best operated as businesses and through corporate approaches to management (Kezar et al., 2019). Individualism is achieved by promulgating values of entrepreneurialism so that workers see themselves as solely responsible for revenue generation and competing with others. Privatization is achieved through market-based values that defund public higher education and encourage a competition for scarce resources.

Values of the individualism/entrepreneurialism, marketization, and privatization of higher education shape academic decision-making and work against goals of equity, and therefore disfavor the advancement of women and racialized minorities (Gill, 2012; Metcalfe and Slaughter, 2008; Veijola and Jokinen, 2018). This rationale can be seen in various changes on campus around employment practices and working conditions, such as the expansion of the contingent labor force. Nearly 70 percent of faculty are not on the tenure track, and women are overrepresented in this group (AAUP, 2020a, 2020b; Flaherty, 2020d). Additionally, this trend also manifests in the "outsourcing" of staff members, who are also predominantly women, and the rising number of postdoctoral scholars, research contingent faculty, and graduate students, now categorized as workers versus apprentices (Kezar et al., 2019). The national average salary for contingent, part-time faculty positions is only $24,000 a year and typically comes without benefits or any other form of crucial support, such as professional development. As women and People of Color move into the ranks of graduate students, postdoctoral researchers, and professors, these positions have become low paid, de-professionalized, and contingent (Kezar et al., 2019).

Under this framework, tenure-track faculty are now incentivized to be entrepreneurial faculty stars who are rewarded for bringing in substantial grants, patents, and licenses. As a result, faculty that engage in service work, mentoring, and student support—typically women and Faculty of Color—are often less successful in tenure-track jobs. Fields that are more difficult to monetize, such as the humanities (also with larger percentages of women), have gone into decline, while growth has been in fields that can be monetized, such as technology, molecular biology, and engineering, which have larger proportions of men (Bérubé and Ruth, 2015).

been made at a variety of institutions unilaterally, resulting in sanctions by the American Association of University Professors.[3]

According to data from the Collaborative on Academic Careers in Higher Education (COACHE),[4] faculty generally rated leadership and academic governance better in late March and April than in September 2020 (Foster, 2020). Faculty gave their administrations this early positive rating because of the quick response of campuses to close down in the face of the COVID-19 pandemic emergency. However, between May and September 2020, faculty have registered concerns of being left out of decision-making processes for months, particularly on decisions that shape domains of teaching and learning, but also more broadly to significant decisions about program closures, finances, and layoffs. For example, one story in *The Chronicle of Higher Education* indicated that Rutgers University had almost no teaching faculty involved in any of the critical decisions made around the COVID-19 pandemic (Taylor, 2020). Indeed, current governance trends appear to be working in opposition to the practices of crisis and equity-minded leadership.[5] In fact, these trends work against practices of effective organizations that typically have a more shared leadership and governance approach that is described later in this chapter (Kezar and Holcombe, 2017).

The gender dynamics observed in the United States across various economic sectors are also apparent in higher education in terms of gender inequality in the workplace (Flaherty, 2020e; Pettit, 2020a). Campus administrators tend to make decisions in a gender-neutral way, which in turn reflects past patterns of inequitable decisions by academic leaders. Because shared governance has been compromised within higher education during the COVID-19 pandemic, many campuses are experiencing a heightening of gender inequalities.

---

[3] For example, in October 2020, Canisius College, Illinois Wesleyan University, Keuka College, Marian University, Medaille College, National University, and Wittenberg University were being investigated by the American Association of University Professors (AAUP) to determine whether these colleges and universities have overstepped their purview and deviated from AAUP's widely followed principles of academic governance during the COVID-19 pandemic, particularly when laying off tenured faculty members (Redden, 2020). AAUP has received dozens of complaints from faculty members about unilateral decisions and actions taken by their governing boards and administrations related to finances, returning to campus, how courses are taught, suspending key institutional regulations, reducing and closing departments and majors, compelling faculty members to teach in person, reducing or cutting payments into retirement plans, and laying off long-serving faculty members (Flaherty, 2020b, 2020f).

[4] COACHE is a database of faculty and academic leaders' views among several hundred college campuses administered out of Harvard University. It is a research-practice partnership between academic institutions and COACHE that administers surveys to participating institutions and provides data to them to support the improvement of faculty work life.

[5] *Crisis leadership* is the process by which an organization deals with a major unpredictable event that threatens to cripple the organization. *Equity-minded* leadership involves being evidence based (i.e., using data), race conscious, institutionally focused, systemically aware, and equity advancing.

Various studies have identified the effects of the COVID-19 pandemic on women faculty's productivity, as discussed in Chapter 3 (Flaherty, 2020a; Viglione, 2020; Vincent Lamarre et al., 2020). Declines in hiring of tenure-track faculty and the increased hiring of women in non-tenure-track faculty positions suggests that fewer women will be available for leadership roles. In addition, during 2020, faculty and academic affairs offices were asked to take the brunt of many budget reductions, and faculty leaders questioned why they have not seen equivalent proposals for downsizing administration and other cost centers (Flaherty, 2020f).[6]

Moreover, full-time faculty faced reduced incomes resulting from furloughs and decreased contributions to retirement programs. Longstanding data show pay disparities between men and women in academe (AAUP, 2017, 2018), so salary inequities are being compounded by the COVID-19 pandemic and associated recession, particularly for households headed by women or single women. One study found that after accounting for academic productivity, regional cost of living, specialty, term length, title, and other factors, women earned $0.12 less than men for every dollar made (Mensah et al., 2020).

It is important for leaders to understand the broad range of issues that should be considered and the many policies that could currently benefit from alteration. Leaders have several resources to support them in these decisions, including work describing appropriate ways to implement tenure clock extension policies, approaches to faculty evaluations to reduce bias toward women and Scholars of Color, forms of support for faculty and their transition to online learning, and acknowledgments and rewards for women, particularly Women of Color, who often take on the majority of service and emotional labor to support students during this difficult time (Gonzales and Griffin, 2020).

DECISION-MAKING DURING THE COVID-19 PANDEMIC

Some emerging data indicate approaches that leaders can use to make decisions, govern, and be accountable in ways that are gender inclusive and help to eradicate growing equity gaps. The predominant approaches include at least three strategies: (1) utilizing the expertise of existing diversity, equity, and inclusion (DEI) staff to inform decision-making processes; (2) creating new structures to address decision-making needs; and (3) altering existing processes to include more voices in decision-making. A few campuses have begun to think about the long-term implications of the COVID-19 pandemic and to recommend strategies to address this issue, such as revising strategic plans aimed at ameliorating equity gaps.

---

[6] Budgetary decisions favoring administrative interests over faculty inherently favor men, who occupy many more administrative positions and the most secure faculty positions.

Utilizing Existing Diversity, Equity, and Inclusion Staff

A case study of the University of Massachusetts Amherst (UMass) exemplifies the approach of capitalizing on existing DEI staff, as well as some other key practices, that can be instructive for other campuses (Clark et al., 2020). Changes UMass made include altering tenure, promotion, and review policies; creating a modified evaluation process highlighting the need for documentation; adapting teaching expectations and evaluations; suspending teaching evaluations; establishing emergency funds for childcare and technology; accommodating salary increases at the time of promotion based on productivity losses; and formally recognizing the intensified caregiving demands. An optional COVID-19 Pandemic Impact Statement was provided for faculty to include in their annual reviews and promotion and tenure cases. This is one of the few campuses to have such a comprehensive array of changes responding to the COVID-19 pandemic and gender inequality issues (Clark et al., 2020).

An analysis of the leadership that produced these changes found that units across the UMass campus sought out the advice of staff participating in the National Science Foundation's ADVANCE program,[7] as well as other staff members on campus with expertise in DEI (Clark et al., 2020). In addition, there was a great deal of coordination across campus units to share ideas about equity recommendations for supporting women faculty and staff. Researchers identified this openness as part of the culture change that ADVANCE had been able to create in terms of a shared commitment and leadership to sustainable equity. The university also had strong senior leadership who spoke out about equity and made it a priority, who met regularly with the ADVANCE team, and who both listened to faculty needs and responded to those needs.

Another effort to include the expertise of existing DEI leaders on campus is illustrated in the letter from chief diversity officers to academic leaders within the University of California system.[8] Campus leaders can also benefit from advice offered by the National Association of Diversity Officers in Higher Education on addressing inequalities during the COVID-19 pandemic.[9]

Creating New Structures

Some institutions have underscored the need for new structures that can support better decision-making and leadership during this critical time. For example, Indiana University is investing in additional racial justice research and is creating a task force to address the negative impact COVID-19 has had on

---

[7] ADVANCE is a program funded by the National Science Foundation to increase the number of and support for women faculty and Faculty of Color in STEM. More information can be found at https://www.nsf.gov/funding/pgm_summ.jsp?pims_id=5383.

[8] Available at https://diversity.universityofcalifornia.edu/policies-guidelines/COVID-19.html.

[9] Available at https://nadohe.memberclicks.net/assets/PressReleases/_NADOHE%20Statement%20on%20DEI%20Training.pdf.

women faculty and researchers. The Gender Equity in Research Task Force at Indiana University explores the negative impact the COVID-19 pandemic has had on research productivity and suggests both short- and long-term actionable solutions within the university's research context. Other leaders have suggested implementing Rapid-Response Leadership Teams that include DEI experts (Goodwin and Mitchneck, 2020), establishing a COVID-19 Pandemic Response Faculty Fellow, and creating a COVID-19 Pandemic Faculty Merit Committee (Flaherty, 2020a). For these new structures, the faculty and administrative leaders designing them are aiming to ensure better decision-making, since this group is specifically tasked with ensuring gender equity, taking a gender advocacy and equity, not neutral, approach.

Altering Existing Processes

A group of concerned faculty at the University of California, Los Angeles (UCLA) suggested that their administration be proactive on and reach out to various existing policy groups, such as tenure and promotion committees, to discuss how to handle the impact of COVID-19. At the same time, they advocated for a new group to develop policies for existing decision groups that may not have this expertise. The faculty informed its administration that it will need guidelines on how to quantify impacts of COVID-19 on teaching, research, and service, as well as clear metrics, tangible benchmarks, and effective communication to decrease bias in merit and promotion decisions (UCLA, 2020). Regarding longer-term accountability and transparency, UCLA faculty leaders are encouraging the university to respond to this COVID-19 pandemic by developing a strategic action plan, which includes metrics and accountability for dealing with changes in faculty productivity because of COVID-19 over the long term.

Other suggestions build on this notion of developing processes for supporting existing institutional decision-making structures but altering these processes so they include different individuals who might be more sensitive to gender equity issues. For example, one group recommended that academic leaders establish inclusive communication, continued monitoring for equitable distribution of resources, and conscientious attention to differential impacts on the workplace climate (Goodwin and Mitchneck, 2020). These investigators also suggested exploring who is at the decision-making table, as it will affect whether gender equity emerges, and engaging campus leaders and experts in DEI, which will broaden participation in decision-making and ensure needed attention to faculty DEI concerns. Another strategy is ensuring funding for DEI work so that this work continues even during the COVID-19 pandemic and inequities do not become larger and more exacerbated. Several efforts are paired with broad-based surveys of faculty and staff to understand specific needs and concerns on a campus related to caregiving, workload, and productivity during the COVID-19 pandemic. (See Chapters 3 and 4 for more information.)

There have also been dozens of suggestions from campuses for a combination of new structures as well as new decision processes that would ensure greater accountability and transparency in decision-making. One instructive example is from the University of Toronto, where a process was proposed for clear internal policies and guidelines aimed at protecting workers. The proposed process included increasing the frequency of open stakeholder meetings to ensure that worker perspectives were considered in decision-making as it pertains to operations during both lockdown and reopening (OCUFA, 2020; University of Toronto, 2020).

LEADERSHIP AND DECISION-MAKING TO ADDRESS CRISES AND INEQUITIES

Based on studies of the type of leadership needed to make equitable decisions and decisions in complex environments such as the COVID-19 pandemic, three key types of leadership—equity-minded, shared, and crisis—can help inform the leadership of administrators, governing boards, and other governing groups and decision-making entities. While a growing body of research provides suggestions about how to create equitable changes in "normal" times, it is particularly important to also look at the literature on crisis leadership to help inform decision-making during the COVID-19 pandemic and similar disruptions. Perhaps surprisingly, some key practices that work during normal times can also work during a crisis when implemented with intentionality.

Equity-Minded Leadership

To reverse the gender inequity trends that have emerged both before and during the COVID-19 pandemic in 2020, leaders in higher education could be well served by taking an equity-minded leadership approach. Equity-mindedness is defined as being evidence based (i.e., using data to explore inequalities), race conscious, institutionally focused, systemically aware, and equity advancing (Dowd and Bensimon, 2014). When practicing equity-mindedness, individuals question their own assumptions, recognize biases and stereotypes that harm the advancement of equity goals, become accountable for closing equity gaps, and see closing racial, gender, and other gaps as their personal and institutional responsibility. To understand and become equity-minded, various practitioners (faculty, administration, staff, etc.) assess and acknowledge that their practices may not be working and understand inequities as a dysfunction of the existing structures, policies, and practices that were not created to serve today's students, and that they can change (Dowd and Bensimon, 2014). A lack of awareness of these issues is what keeps reproducing gender inequalities over time. While there is a need to continue to change the representation of leaders so that they have greater awareness of different circumstances, such as race and gender, equity-mindedness focuses on leaders of any background being able to adopt an equity mindset (Kezar and Posselt, 2020). Since the race and gender of current leaders is unlikely to change soon, equity-mindedness is particularly important at this time to make needed changes.

Equity-minded leaders can have both immediate and lasting impacts on a campus's ability to close equity gaps and goals (Galloway and Ishimaru, 2015; Kezar and Posselt, 2020; Shields, 2010; Theoharis, 2007). For example, research documents how leaders that adopt an equity-minded approach have been successful in closing equity gaps for students in college (Dowd and Bensimon, 2014). Equity-minded leaders dismantle discriminatory policies, use data and assessment to understand inequity, and shift the consciousness among educators when it comes to discrimination and bias (ASCCC, 2010; CDE, 2010; Felix et al., 2015; Galloway and Ishimaru, 2015; Santamaría, 2014). Several of the suggested structures offered by campuses, such as rapid response teams, COVID-19 pandemic task forces, highlight the importance of having a mechanism for integrating equity-minded thinking into decision-making processes at campuses (Goodwin and Mitchneck, 2020). While they were forming, it might have been helpful for these rapid-response and COVID-19 pandemic teams to look at the equity-minded leadership literature to help support their work.

Shared Leadership

Shared leadership is "the dynamic, interactive influence process among individuals in groups for which the objective is to lead one another to the achievement of group or organizational goals or both" (Pearce and Conger, 2003).[10] A review of literature on shared leadership found four key elements that characterize shared leadership (Kezar and Holcombe, 2017):

- A greater number of individuals take on leadership roles than in traditional models.
- Leaders and followers are seen as interchangeable. In some cases, this may mean that leadership occurs on a flexible and emergent basis, while in others it rotates more formally.
- Leadership is not based on position or authority. Rather, individuals with the expertise and skills needed for solving the problem at hand are those that lead. To that end, multiple perspectives and expertise are capitalized on for problem solving, innovation, and change.
- Collaboration and interactions across the organization are typically emphasized.

Inherent in this approach is a greater honoring of the multiple perspectives that make up a campus, which typically leads to much more inclusive

---

[10] This section draws largely from Kezar and Holcombe (2017).

decision-making and equitable outcomes. Decentralization and the promotion of local autonomy increase the adaptability of organizations and allow them to respond creatively and quickly to changing environmental conditions (Heifetz, 1994; Wheatley, 1999).

Studies of shared leadership demonstrate that it tends to develop decisions that are more inclusive and equitable and to represent more diverse perspectives (Pearce and Conger, 2003; Wheatley, 1999). Studies have explored the potential of shared leadership for improving gender equity and found that it is associated with better performance for students and creating a better work environment for teachers and administrators (Hrabowski III, 2019; Shakeshaft et al., 2007). Many of these studies suggest that shared leadership is especially beneficial in complex environments that require frequent adaptations, such as a pandemic (Feyerherm, 1994; Pearce and Sims Jr., 2002; Pearce et al., 2004).

Crisis Leadership

There have been several studies specifically focused on crisis leadership in higher education (Fernandez and Shaw, 2020; Gigliotti, 2019, 2020) that have identified three leadership practices to help navigate a crisis: (1) connecting with people broadly as individuals and establishing mutual trust, (2) distributing leadership throughout the organization, and (3) communicating clearly and often with all stakeholders (Fernandez and Shaw, 2020). Reports in the academic trade press suggest that faculty and staff have been advocating for regular meetings with senior university leaders and for creating avenues for communication between decision makers and those affected by the decisions (Flaherty, 2020b, 2020c, 2020f).

Effective leadership during a crisis also benefits from shared or distributed leadership so that those with expertise about policies and practices at the ground level can easily communicate with those who have decision-making authority. During a crisis, leaders who emphasize empowerment, involvement, and collaboration allow themselves a greater degree of agility and innovation than is possible with an inflexible hierarchical leadership paradigm (Kezar and Holcombe, 2017).

The third area of consensus related to crisis leadership is clear, frequent communication with stakeholders. During a crisis, it is important to use multiple communication channels (Robbins and Judge, 2018). In the context of the COVID-19 pandemic, when people are unable to engage in face-to-face communication because of social-distancing practices, leaders considered live streaming of updates or messages of encouragement (Fernandez and Shaw, 2020). The choice of communication approach should also consider stakeholders' preferences. Faculty and staff may prefer updates from leadership through email, while students may prefer a variety of social media platforms or text messages.

Researchers have studied campuses in crisis and identified active listening as another area of communication important for quality decisions (Kezar et al., 2018). Active listening is a structured form of listening and responding that focuses the attention on the speaker—instead of on one's own perspectives—and improves mutual understanding without debate or judgment (Kezar et al., 2018). Many of the emerging recommendations from faculty and staff during the COVID-19 pandemic have also related to more transparency with decision-making and increased communication. From accounts in the media, the current communication approach on campuses is failing and is exacerbating inequities (Flaherty, 2020b, 2020f).

DATA GAPS ON ACADEMIC LEADERSHIP AND DECISION-MAKING

There are limitations in the literature on academic leadership and decision-making. It has been almost a decade since the National Center for Education Statistics stopped collecting data for the National Study of Postsecondary Faculty, resulting in limited data about decision-making and leadership related to the COVID-19 pandemic. In addition, studies of successful leadership practices during the COVID-19 pandemic in 2020 were often based on single case studies and should be understood to have limited ability to generalize. For example, the COACHE faculty job satisfaction survey is one of the only sources of faculty members' views on governance and leadership. There has been intermittent data collection related to leadership representation in higher education with a focus on presidents and governing board members, which leaves critical gaps in knowledge related to other administrative roles. Similarly, there are not enough data collected on Leaders of Color or studies that evaluate the impact of decision-making on gender as well as racial and ethnic inequalities. Few studies have incorporated an intersectional perspective, grounding race and evaluating the intersection with other identities such as gender (Harris and Patton, 2019).

CONCLUSION

Historically, gender inequalities are pervasive among leaders in higher education leadership and their decision-making process, and the emergence of the Gig Academy has exacerbated gender inequalities. However, many campuses have existing efforts or offices (e.g., ADVANCE, DEI offices) working on changing the campus cultures and support equity over the long term. Leaders who are focused on addressing gender equity would benefit from working strategically to address the emerging gender inequalities of the COVID-19 pandemic observed during 2020. Many campuses may need more mechanisms for bringing faculty and administrators together around decision-making and leadership, particularly for non-tenure-track faculty that are often excluded from governance. While there is evidence from the first 9 months of the COVID-19 pandemic of worsening gender inequalities, there are also existing strategies available for reversing these trends.

7

Mental Health and Well-being[1]

INTRODUCTION

During normal times, being disadvantaged has been associated with worse mental and physical health (Vanderbilt et al., 2013; Williams, 2018). However, during times of extreme societal stress, such as during 2020, the negative effects of being disadvantaged are often accentuated. For example, according to the Bureau of Labor Statistics, the number of unemployed people in the United States increased by 6.8 million between February 2020 and October 2020, with women, particularly those who identify as Black or Latina or disabled, affected disproportionately. In addition, as mentioned previously, COVID-19 infection and death rates are higher among those from socioeconomic or race/ethnic minority groups; this is also true for individuals with underlying health conditions, including mental illness (Boserup et al., 2020; Fond et al., 2020).

This chapter focuses on the mental health effects of the COVID-19 pandemic for women in science, technology, engineering, mathematics, and medicine (STEMM). The evidence presented here highlights ways mental health may affect women's engagement in STEMM fields. In particular, this chapter provides evidence that psychosocial, professional, and biological factors contribute to greater risk for mental health concerns among academic women versus men in STEMM fields.

The evidence available at the end of 2020 from across the globe indicated that women in the general population, particularly those on the frontlines of

---

[1] This chapter is primarily based on the commissioned paper "The Impact of COVID-19 on the Mental Health of Women in STEMM," by C. Neill Epperson, Elizabeth Harry, Judith G. Regensteiner, and Angie Ribera.

health care, were at greatest risk of adverse mental health effects during the COVID-19 pandemic. There were, however, few studies focusing on the effects of stress on, or the mental health of, women in STEMM fields outside of medicine and nursing. Therefore, this chapter discusses key indicators and evidence-based assessments for burnout and mental illness, gender differences in stress exposures, the effects of epidemics and pandemics on the workforce in specific STEMM fields such as medicine and nursing, and the effects of social-distancing public health requirements on university students and faculty more generally, as well as interventions to promote well-being among academic women in STEMM fields. Where appropriate, data from studies focusing on the impact of previous epidemics and the COVID-19 pandemic on the mental health of women in health care and those in the general population were extrapolated to women in academic STEMM. Finally, the majority of the data reviewed assumes that women are those who were identified as women at birth (cisgender women). Where it is possible, this chapter describes how the intersectionality of gender minority status may be affected by the current crisis. Similarly, the chapter utilizes data from the general population to extrapolate to People of Color, with a few exceptions noted.

EFFECTS OF ISOLATION AND SOCIETAL STRESS FOR WOMEN IN STEMM

It is well documented in both preclinical and human studies that chronic and unpredictable stress, such as what occurred during the COVID-19 pandemic in 2020, is the most detrimental form of stress for health (Yaribeygi et al., 2017). Gender differences in stress exposures and in biological response to stress may interact to increase risk of mental health problems for women during the COVID-19 pandemic. Factors such as social isolation, caregiving, and job insecurity, all more common among women during previous pandemics, have been associated with greater mental health concerns (Connor et al., 2020).

Discrimination and marginalization have long been recognized as stressors and contributing factors to poor mental health (Schmitt et al., 2014; Sutter and Perrin, 2016). Women and other underrepresented groups studying and working in STEMM fields are disadvantaged to a greater degree than their male counterparts (Myers et al., 2020; Woitowich et al., 2020). Women in academic STEMM fields are more likely to be early in their career (NASEM, 2020), have a lower salary regardless of professional ranking in STEMM (Raj et al., 2019), be a single parent or a primary caregiver (Calisi et al., 2018; Jolly et al., 2014; Yavorsky et al., 2015), and report experiencing greater work-related stress (Ornek and Esin, 2020) and discrimination in the workplace or their community (Jagsi et al., 2016; Lu et al., 2020). Stressors such as these (discussed more fully in Chapters 2, 3, and 4) are compounded for women by the same social isolation, work disruption, financial worries, and health concerns experienced by others during the COVID-19 pandemic.

Social isolation can be an additional stressor specifically for women in STEMM. Social support—particularly that gained from in-person contact—is a protective factor against the adverse effects of stress on health (Connor et al., 2020), and during many recent societal stressors in the United States, such as natural disasters and terrorist attacks, individuals have been able to gather with family, friends, and colleagues to grieve and heal. Epidemics and pandemics are unique in their requirement for social distancing, which is in direct opposition to human nature under times of stress and increases risk for poor mental and physical health (Umberson and Montez, 2010).

Particularly in fields such as engineering, physics, computer science, and certain subspecialties of medicine, women are likely to be in the minority and have fewer women role models at the rank of professor or in other leadership positions (NASEM, 2020). Because women are more likely than men to use social relationships to cope with stress or threat (Smith, 2014; Taylor et al., 2000), social distancing during the COVID-19 pandemic could exacerbate the relative lack of social support from women colleagues, mentors, and role models, as discussed in Chapter 5. For example, women university students who use a coping style characterized by greater social supports showed a reduction in physiologic response to stress, both across the day and during a laboratory stressor (Sladek et al., 2017). Taken together with exposure to fewer women in the workplace and the importance of social support to stress regulation among women, it may be beneficial for leaders in academic institutions to consider the social distancing required during the COVID-19 pandemic as they create programs to maintain engagement of academic women in STEMM fields (see Chapter 6 for more on leadership).

Finally, during pre-COVID-19 pandemic times, women across the globe were more likely to have depression, anxiety, posttraumatic stress, and insomnia (Bracke et al., 2020). The main mental health conditions most exacerbated by recent societal stressors—such as terrorist attacks, natural disasters, and infectious disease outbreaks—are insomnia, depression, anxiety, posttraumatic stress, and alcohol and drug use (Cabarkapa et al., 2020; Esterwood and Saeed, 2020). Each of these, with the exception of alcohol and drug abuse, are disorders that occur more frequently among women (Bracke et al., 2020).

EFFECTS OF COVID-19 PANDEMIC-RELATED STRESS ON WOMEN IN STEMM

Several organizations, including the Centers for Disease Control and Prevention, Veterans Health Administration, state governments, and public health agencies, have developed web pages listing a wide range of COVID-19–associated sources of stress, such as personal, family, and community health related to the risk of infection; financial related to loss of job or wages; childcare resulting from school and/or daycare closures; social isolation; and the uncertain future trajectory of the COVID-19 pandemic and its consequences (Park et al., 2020; UN, 2020). These kinds of stressors may affect women and men differently. As discussed in Chapter 4, women tend to be the major caregivers within an extended family, and, as a result, they are more likely to experience increased stress associated with caring for themselves, loved ones, or friends who contract disease. Along with social isolation and sheltering, there has been an increase in domestic violence during the COVID-19 pandemic, adding yet another stressor predominantly for women (Boserup et al., 2020).

Women are at greater risk of medical issues such as endocrine, immune, rheumatologic, and neurologic conditions that are frequently comorbid with depression and anxiety (Desai and Brinton, 2019; Golden and Voskuhl, 2017; van der Woude and van der Helm-van Mil, 2018). Many of these conditions are stress sensitive, increasing the risk of an exacerbation during the COVID-19 pandemic (Gazerani and Cairns, 2020). In addition, women are more likely to be diagnosed with autoimmune diseases that put them into an at-risk category that could affect their ability to work in any STEMM field that requires some level of social or in-person contact during the COVID-19 pandemic.

There are special concerns related to women health-care workers. As with any pandemic, frontline health-care workers are most at risk for exposure and contracting SARS-CoV-2.[2] However, approximately 77 percent of the health-care frontline workforce is made up of women, creating a greater overall infection risk for women in this STEMM field (Robertson and Gebeloff, 2020). In addition to caring for their own families (see Chapters 2 and 4 for more information), women health-care workers are more likely than men health-care workers to be at the bedside taking care of patients with COVID-19 and managing the distress of family members and of the sick and dying (Lai et al., 2020). Women health-care workers are also more likely than men health-care workers to work shift-based schedules that can be unpredictable and can negatively affect circadian rhythms and sleep (Lai et al., 2020).

Mothers-to-be in STEMM, meanwhile, face additional COVID-19 pandemic stressors and consequences (Staniscuaski et al., 2020). Prenatal care, delivery, and infant needs present financial challenges and stress (Ahlers-Schmidt et al., 2020), and concerns about infection during pregnancy not only create stress but lead to avoidance of medical services and worries about how delivery will occur (Berthelot et al., 2020; Preis et al., 2020a, 2020b). Given that increased maternal stress will negatively affect the progression of pregnancy, it is imperative for all mothers to receive care that minimizes disease exposure and delivery complications. Similar to women in STEMM with underlying medical conditions, pregnant women in STEMM have to consider the location of their work and whether colleagues and labmates are careful about their own exposures to the

---

[2] The virus that causes COVID-19.

novel coronavirus. Institutional procedures to ensure safety in the workplace by ensuring access to testing, screening those who come to campus, and rapidly responding to contain outbreaks are essential for these women to be able to use laboratory facilities.

For women in academic STEMM in general, the COVID-19 pandemic has exacerbated many stresses women in academia face under usual conditions, as discussed in Chapters 3, 4, and 5 (Howe-Walsh and Turnbull, 2016). For example, the mantra "publish or perish" emphasizes survival, let alone success, in academia and requires one to continually publish papers and obtain funding. As described more fully in Chapter 3, the effects of the COVID-19 pandemic on women in academic STEMM fields has already been observed as a decrease in productivity. With variation by discipline, women published fewer papers and received fewer citations of their work between March 2020 and December 2020 (Amano-Patino et al., 2020; Andersen et al., 2020; Gabster et al., 2020).

THE COVID-19 PANDEMIC AND THE MENTAL AND PHYSICAL HEALTH OF WOMEN IN STEMM

During 2020, the COVID-19 pandemic accentuated gender differences in mental health concerns such as depression, anxiety, posttraumatic stress, and insomnia (Carmassi et al., 2020; Guadagni et al., 2020; Pappa et al., 2020). Delays in clearance for conducting research during 2020, a result of the COVID-19 pandemic, led researchers to experience increased burnout, sleep disturbance, poor appetite, increased interpersonal problems, and decreased motivation (Sharma et al., 2020). The confluence of major events, including the COVID-19 pandemic, racial injustices, and geopolitical unrest, affected academic faculty in multiple domains professionally and personally (Gruber et al., 2020). Each of these outcomes serves as a surrogate measure of well-being and risk for mental health problems during and after the COVID-19 pandemic.

There is a reciprocal relationship between employee well-being and institutional success (Attridge, 2007, 2009).[3] Employee well-being affects institutional metrics and culture, while institutional culture, policies, and procedures affect individual employee well-being. This section discusses key indicators that leaders can use to identify risk of declining employee well-being, including mental illness, burnout, and sleep disturbance.

---

[3] *Employee well-being* is defined "as an integrative concept that characterizes quality of life with respect to an individual's health and work-related environmental, organizational, and psychosocial factors. Well-being is the experience of positive perceptions and the presence of constructive conditions at work and beyond that enables workers to thrive and achieve their full potential" (Chari et al., 2018, p. 590).

Burnout as a Key Indicator of Overall Mental Well-being and Job Satisfaction

Burnout can be measured and has documented negative effects on individuals in the workforce, with considerable attention paid to individuals in health-care professions (NASEM, 2019b).[4] Academic medical centers pay particular attention to the prevalence and prevention of burnout among health-care workers, including the effects of burnout on patient safety, quality of care, and professionalism (Panagioti et al., 2018). Across the workforce more generally, high levels of burnout (variably defined) have been associated with a number of somatic conditions, including high blood pressure, coronary artery disease, and diabetes (Guan et al., 2017; von Känel et al., 2020). Women in medicine, nursing, and basic science research report higher levels of personal and work-related burnout than men in similar roles (Gold et al., 2016; Linzer et al., 2000; Messias et al., 2019; Rabatin et al., 2016). These gender differences in the individuals' relationship to work starts early in academic training. A recent 3-year longitudinal study from Germany comparing freshmen medical students with STEM students indicated that STEM students started and continued to demonstrate greater burnout-related risk patterns compared with medical students. Women students showed a more unfavorable pattern regardless of group (Voltmer et al., 2019).

Institutional leaders have several validated tools to measure burnout available to them, including the Maslach Burnout Inventory–Human Services Survey for Medical Personnel and the Copenhagen Burnout Inventory (designed to be used for any occupation; see Table 7-1; NAM, n.d.). Similarly, there are several validated tools to measure composite well-being, including the Stanford Professional Fulfillment Index. These tools can be used to monitor burnout and well-being among those in the STEMM environments, with particular attention to vulnerable populations such as women, trainees, and people with identities historically marginalized or excluded in STEMM.

Factors Contributing to Burnout

During 2020, the COVID-19 pandemic exacerbated many of the long-standing factors that contribute to greater burnout among women, compared with men, in the STEMM professions. Prior to the COVID-19 pandemic, women in STEMM reported greater emotional exhaustion, a domain of burnout; greater cynicism; and lower academic efficiency in environments described as "chilly" and unwelcoming to women (Jensen and Deemer, 2019). A comprehensive review

---

[4] The World Health Organization defines *burnout* as "a syndrome conceptualized as resulting from chronic workplace stress that has not been successfully managed. It is characterized by three dimensions: (1) feelings of energy depletion or exhaustion; (2) increased mental distance from one's job, or feelings of negativism or cynicism related to one's job; and (3) reduced professional efficacy. Burn-out refers specifically to phenomena in the occupational context and should not be applied to describe experiences in other areas of life" (WHO, 2019).

of burnout among university teachers working in multiple countries indicated that men are more likely to report cynical and negative approaches to others (depersonalization), while women were more likely to report greater emotional exhaustion (Purvanova and Muros, 2010; Wyatt and Robertson, 2011), both symptoms of burnout (Maslach and Jackson, 1993). Moreover, there is some research indicating that women were more likely to score higher than men on all dimensions of burnout (Wyatt and Robertson, 2011). Being younger, having a larger student load, and growing tuition costs among students were also associated with greater symptoms of burnout.

When juggling more domestic responsibilities, as discussed in Chapters 2 and 4, women experience a higher overall cognitive load, putting them at higher risk of burnout. For example, when women physicians are performing more of the domestic responsibilities, they are more likely to wish for a career change, particularly when in a procedural field (Lyu et al., 2019). In addition, high task loads of workplace environments may also contribute to enhanced burnout (Harry et al., 2019). Women are also more likely to take time at home or reduce hours to accommodate the COVID-19 pandemic, as noted in Chapters 3 and 4, which may propagate gender inequities and gaps in compensation (Brubaker, 2020). Leaders in STEMM environments can take steps to measure the climate for women in their institution, have a process to monitor if more women than men are decreasing hours or full-time equivalents, and evaluate the task load placed on employees using a validated tool such as the National Aeronautics and Space Administration (NASA) Task Load Index (NASA, n.d.).

Sleep Quality

Insomnia is both a symptom and a predictor of onset or exacerbation of a number of mental health disorders, including depression, bipolar disorder, post-traumatic stress disorder (PTSD), and substance abuse. Sleep disturbance is also associated with greater risk for suicide among those with mental illness (R.T. Liu et al., 2020; Weber et al., 2020). Prior to the COVID-19 pandemic, studies examining sleep have focused on the workforce in general or specific patient or demographic populations, with health-care workers, particularly nurses, reporting worse sleep quality than the general population (Khatony et al., 2020; Zeng et al., 2019).

Psychological stress is a primary contributor to reductions in sleep quality (Kim and Dimsdale, 2007), and many of the COVID-19 pandemic-related stressors create risk for the onset and worsening of insomnia and health burdens related to poor sleep quality among frontline health-care providers (Kobayashi and Mellman, 2012). Since the COVID-19 pandemic and its associated public health measure were implemented across nations, a focus has been on health-care workers and students. Similar to reports with previous coronavirus epidemics, such as severe acute respiratory syndrome, or SARS, and Middle East respiratory syndrome, or MERS, insomnia is one of the most common and consistently reported concerns among health-care workers (Pappa et al., 2020) and is a predictor of job exit among middle-aged and older adults (Dong et al., 2017).

Data from multiple nations indicate an increase in poor sleep quality and complaints of insomnia in the general population during the COVID-19 pandemic, but to a significantly greater degree among health-care workers (Cabarkapa et al., 2020; Li et al., 2016; Marelli et al., 2020; Romero-Blanco et al., 2020; Sheraton et al., 2020). For example, between March 1 and April 30, 2020, health-care workers in Spain reported a higher prevalence of new onset or worsening of insomnia compared with non-health-care workers (Herrero San Martin et al., 2020). Given many of the stressors experienced by women health-care workers are also experienced by women in academic STEMM fields, much of these data can potentially be generalized.

Sleep quality among university students may also be unduly affected by the shutdown. A recent longitudinal (pre- and post-lockdown) study of 207 nursing students in Spain revealed that being a woman, being a first- or second-year student, living with one's family, and use of alcohol were associated with significantly worse sleep quality (Romero-Blanco et al., 2020). Similarly, a longitudinal study including Italian university students indicated a worsening in sleep parameters, particularly among women students, during the lockdown from March 10, 2020, to a partial lifting on May 3, 2020 (Marelli et al., 2020). While more than half of the students reported clinically meaningful sleep problems before the COVID-19 pandemic shutdown, almost three-quarters fell into this range after the shutdown was partially lifted.

Women who have caregiving responsibilities also report changes to sleep quality during the COVID-19 pandemic. A recent study (Zreik et al., 2020) in Israel found an increase of self-reported insomnia in mothers with at least one child between the ages of 6 and 72 months during home confinement related to the COVID-19 pandemic from 11 percent (retrospective report) to 23 percent. Women who are balancing work-related expectations with "sandwiched care" responsibilities (such as caring for both emerging adult children and elderly parents; see Chapter 4) can greatly impact sleep quality and may lead to women leaving academic STEMM (Mavriplis et al., 2010). Importantly, insomnia was found to be a predictor of employed workers ages 50–70 leaving the workforce because of poor health (Dong et al., 2017). Hence, institutions cannot afford to ignore insomnia as a potential contributing factor to women leaving academic STEMM during and after the COVID-19 pandemic.

Relevant sleep quality for People of Color, everyday discrimination, and the perceived stress of racism are associated with worse self-reported sleep quality (Grandner et al., 2012; Vaghela and Sutin, 2016). As racism and other forms of discrimination in the workplace remain common (Fekedulegn et al., 2019), particularly for Women of Color, it would be expected that Black women in STEMM carry the additional burden of perceived stress of racism on sleep quality during COVID-19. Moreover, it is well documented that the physical and mental health of People of Color and their access to medical interventions, particularly for mental illness, were lacking even prior to the COVID-19 pandemic (Breslau et al., 2017; IOM, 2011; Mangrio and Sjögren Forss, 2017). Given that mental health conditions are also associated with a decreased utilization of primary care services (DeCoux, 2005), the intersection of COVID-19 pandemic-related stress, reductions in access to medical care, fear of infection, and other barriers may have a greater impact on women and People of Color in the months and years to come. However, the relationship between the effects of mental health and chronic health conditions for women compared with men in the setting of the COVID-19 pandemic was not understood at the end of 2020.

Assessing Mental Health in the STEMM Workforce

Validated tools such as the 2-item and 9-item Patient Health Questionnaires, or PHQ-2 and PHQ-9 (Kroenke et al., 2001, 2003), the Generalized Anxiety Disorder 7-Item (GAD-7; Spitzer et al., 2006) questionnaire, the Posttraumatic Stress Disorder Checklist (PCL; Bovin et al., 2016), and sleep-related impairment measures, such as the Insomnia Severity Index or the Pittsburgh Sleep Quality Index (Buysse et al., 1989), are validated measures to assess the likelihood of developing a serious mental condition (Morin et al., 2011; NAM, n.d.; Spitzer et al., 2006; Yu et al., 2011). A description of these instruments and symptom severity range is included in Table 7-1. Given continued concerns among the academic and medical workforce that seeking mental health care will adversely affect their professional standing among their peers and threaten their career opportunities (Feist et al., 2020), institutions may experience resistance to widespread assessment for mental health conditions. Utilizing burnout scales or assessments for sleep disturbances may be a less stigmatizing method to obtain a proxy for faculty mental health during and after the pandemic.

COVID-19–Specific Stress Measures

Essential workers in STEMM fields, such as medicine and nursing, work at the epicenter of the COVID-19 pandemic and, as a result, face increased risk for infection and overall higher rates of stress. The recently developed COVID Stress Scales (CSS; Taylor et al., 2020) categorize stressors from the COVID-19 pandemic into five categories: danger and contamination fear, social and economic stress, traumatic stress symptoms, checking and reassurance-seeking behavior, and xenophobia. Recent findings suggest that the five factors of the CSS form a COVID-19 pandemic stress syndrome. In the general population, each of these factors can contribute to increased substance use and abuse risk. These factors can be compounded in essential workers, placing this group at particularly high risk for substance use and abuse (McKay and Asmundson, 2020). While there were no data available at the end of 2020 specifically for women and People of Color

TABLE 7-1 Validated Measurements of Mental Well-being

Instrument	Description
Patient Health Questionnaire – 2-Item (PHQ-2; Kroenke et al., 2003)	Assesses depressed mood and decreased interest/pleasure. Responses ranging from 0 = *not at all* to 3 = *nearly every day*. Total scores of 3 are considered indicative of major depression. Reference period = past week.
Patient Health Questionnaire – 9-Item (PHQ-9; Kroenke et al., 2001)	Frequency of depression symptoms are measured using a scale ranging from 0 = *not at all* to 3 = *nearly every day*. Scores range from 0 to 27 with mild (5–9), moderate (10–14), moderately severe (15–19), and severe (20–27) depression. Reference period = past week.
Generalized Anxiety Disorder – 7-Item (GAD-7; Spitzer et al., 2006)	Measure of generalized anxiety disorder. Frequency of anxiety symptoms are rated as 0 = *not at all* to 3 = *nearly every day*. Total score range = 0 to 21, with mild (5–9), moderate (10–14), and severe (15–21) anxiety. Reference period = past 2 weeks.
Insomnia Severity Index (ISI; Morin et al., 2011) (7 items)	Items 1–3 assessed the nature of insomnia with questions related to problems falling asleep, staying asleep, and early awakening (0 = *no problem* to 4 = *very severe*). Item 4 assessed dissatisfaction with sleep (0 = *very satisfied* to 4 = *very dissatisfied*). Items 5–7 assessed the impact of insomnia by asking about sleep difficulties interfering with daytime functioning, etc. (0 = *not at all* to 4 = *very much*). Scores range from 0 to 28 with >10 considered indicative of insomnia.
Pittsburgh Sleep Quality Index (PSQI; Buysse et al., 1989) (19 items)	Individuals report on seven components of sleep: sleep quality, sleep latency, sleep duration, habitual sleep efficiency, sleep disturbances, use of sleeping medication, and daytime dysfunction. Scores on the seven components (weighted equally on a 0–3 scale) are added for a total score of 0–21. A score of >5 is considered indicative of poor sleep quality. Reference period = past month.
PTSD Checklist for DSM-5 (PCL-5; Blevins et al., 2015) (20 items)	Respondents report on how much they have been bothered by each PTSD symptom using a scale ranging from 0 = *not at all* to 4 = *extremely*. The items tap into four subscales: re-experiencing, avoidance, hyperarousal, and negative alterations in cognition and mood. Total score ranges from 0 to 80. Reference period = past month.
Maslach Burnout Inventory – Human Services Survey for Medical Personnel (MBI-HSS MP; Maslach and Jackson, 1993) (22 items)	Measures three aspects of burnout: emotional exhaustion (EE; 9 items), depersonalization (DP; 5 items), and low sense of personal accomplishment (PA; 8 items). Frequency of experiences ranges from 0 = *never* to 6 = *every day*. Items are added for a total score ranging from 0 to 54 for EE, 0 to 30 for DP, and 0 to 48 for PA. A score of 27 on the EE subscale and a score of either 10 on the DP subscale or 33 on the PA subscale are considered indicative of burnout.
Stanford Professional Fulfillment Index (PFI; Trockel et al., 2018) (16 items)	Measures burnout and personal fulfillment in physicians. Four items assess work exhaustion (e.g., EE at work) and 6 items assess interpersonal disengagement using a scale ranging from 0 = *not at all* to 4 = *extremely*. The 6 items related to professional fulfillment (e.g., my work is meaningful to me) use a scale ranging from 0 = *not at all true* to 4 = *completely true*. The items are added for a total score of 0 to 64. Reference period = past 2 weeks.

TABLE 7-1 Continued

Instrument	Description
Copenhagen Burnout Inventory (CBI; Kristensen et al., 2005) (19 items)	Measure of burnout in any occupational group. Three aspects of burnout are assessed: Prolonged physical or psychological exhaustion perceived to be related to personal (6 items) or work life (7 items) and 6 items related to working with clients. Items are rated on frequency where 0 = *never or almost never*, 25 = *seldom*, 50 = *sometimes*, 75 = *often*, 100 = *always* or 0 = *to a very low degree;* or bother, where 25 = *to a low degree*, 50 = *somewhat*, 75 = *to a high degree*, and 100 = *to a very high degree*. Subscale range is 0–100.
NASA Task Load Index (NASA-TLX; Hart and Staveland, 1988) (6 items)	Assesses subjective experience of workload. Individuals report on six dimensions: mental demand, physical demand, temporal demand (i.e., time pressure to complete tasks), performance, effort, and frustration level. Each dimension is rated on a 0 to 100 scale in 5-point increments. The dimensions can be weighted by using 15 pair-wise comparisons of the dimensions (e.g., comparing whether mental demand versus physical demand contributed more to workload). Each dimension can be chosen from 0 (not relevant) to 5 (more important than any other dimension) times. Ratings of dimensions deemed to be most important in creating the workload of a task are given more weight in computing an overall workload score.

regarding scores on the CSS, one can extrapolate that any group that came into the COVID-19 pandemic facing health disparities or lower professional standing (e.g., lack of seniority and lower salary) could experience the COVID-19 pandemic as a greater threat to their health and financial stability.

Depression and Anxiety

Depression and anxiety are frequently comorbid conditions, and most studies of mental well-being during the COVID-19 pandemic have measured both, often using standardized ratings such as the PHQ2/PHQ9 and the GAD-7 (see Table 7-1). As measured in 2020, the prevalence of depression symptoms, including those in the moderate to severe range, in the general U.S. population was 3-fold higher during the COVID-19 pandemic than before (Ettman et al., 2020). The risk factors included lower economic resources and greater exposures to stressors.[5] Literature that focuses on mental health among health-care workers also consistently reports being a woman as a risk factor for adverse mental health consequences of the COVID-19 pandemic.

The earliest results came from China, where SARS-CoV-2 first spread to humans. A cross-sectional, web-based study conducted in February 2020 showed

---

[5] Findings were not reported by sex or gender nor type of employment, making it difficult to examine these data in relationship to the goal of understanding the effects of the COVID-19 pandemic on women in STEMM.

that overall psychological problems, including depression, anxiety, and insomnia, were reported by a majority of physicians, medical residents, nurses, technicians, and public health professionals. Being a frontline woman health-care worker was associated with greatest risk for both anxiety and depression (Que et al., 2020). These data have been borne out in a meta-analyses and scoping reviews of studies focusing on mental health during the COVID-19 pandemic (Krishnamoorthy et al., 2020; Shaukat et al., 2020). One group (Linos et al., 2020) studied anxiety among U.S. physician mothers using the GAD-7 survey and found that 41 percent of 1,809 participants scored as having moderate or severe anxiety. Being a shift worker, being a nurse, caring for infected patients, and being a woman are each most consistently reported as risk factors for depression and anxiety. The finding of vulnerability to depression and anxiety among women during the COVID-19 pandemic is similar to that reported for previous viral epidemics, including MERS, Ebola, and SARS (Cabarkapa et al., 2020).

These findings in health-care workers extend to trainees. Trainees in health care who are exposed to patients with COVID-19 report significantly higher stress than trainees not caring for these patients (Kannampallil et al., 2020). This study also found that women trainees were more likely to be stressed regardless of patient exposure status, while unmarried trainees were significantly more likely to be depressed and marginally more likely to have anxiety.

Prior to the pandemic, more than one-third of graduate students reported seeking mental health care resulting from the stress of their studies and uncertainty about their career (Woolston, 2019). A survey of graduate students during the COVID-19 pandemic indicates that depression is equally high for women and men, but women are more likely to report symptoms of anxiety than are men. Symptoms of psychological distress were even higher among Latinx students and students who identify as lesbian, gay, or bisexual. Depression was most common among students in the physical sciences, and anxiety was more common among those in biomedical research. Of the 5 percent of students who reported not adapting well to online learning, almost two-thirds reported high anxiety levels (Woolston, 2020d). With the added stress of having to study virtually in many cases, having reduced laboratory access, and growing concerns about federal grant funding, particularly for topics that are not related to the COVID-19 pandemic, it follows that graduate students will likely require more mental health care as a result of the COVID-19 pandemic (Woolston, 2020e).

Trauma Exposures and Posttraumatic Stress Symptoms

Within 1 month of the COVID-19 epidemic emerging in Wuhan province, investigators in China examined posttraumatic stress syndrome (PTSS) symptoms using the PTSD Checklist, specifically the PCL-5 (Bovin et al., 2016), and sleep quality using the Pittsburgh Sleep Quality Index among current or recent residents of Wuhan. Women reported more PTSS symptoms than did men. Women also experienced higher degrees of reexperiencing symptoms, negative alterations in cognition or mood, and hyperarousal compared with men (N. Liu et al., 2020). Better subjective sleep scores were associated with lower PCL-5 scores. Treating both the PTSS symptoms as well as the sleep disturbance is critical to recovery, as poor sleep quality has been linked to the onset and maintenance of PTSD (Richards et al., 2013, 2020).

While evidence indicates that women in STEMM on the frontline of the COVID-19 pandemic are at greater risk or poor mental health, a review of data from previous pandemics and early studies of health-care workers from the COVID-19 pandemic provides a glimpse into risk and resilience factors for PTSD and PTSS (Carmassi et al., 2020) (see Table 7-2). One can also extrapolate from these data in health-care workers to consider women in other academic STEMM fields. For example, unpredictability at work and having to learn new strategies to accomplish one's work are risk factors for PTSS in the context of a pandemic. Cognitive overload is also a risk factor, particularly in the face of stress and trauma. Providing clarity regarding safety measures and making sure that individuals feel adequately trained to meet the needs of their job during the pandemic are critical for mental health. Social supports and personal traits such as

TABLE 7-2 Risk and Resilience Factors: Documentation from Health-Care Workers Extrapolated to Academic Women in STEMM

	Health-Care Worker		Women in STEMM Fields
Risk Factors			
Unpredictability at work	Daily caseload		Access to laboratory and equipment
Managing expectations	Families and patients		Mentors, program officers
Increased acuity	Increase in critically ill		Meeting deadlines
Decision-making burden	Increased executive load		Increased executive load
Traumatic exposures	More deaths per day		Threat of infection, loss of job
New procedures	New treatments to learn		"COVID-izing" research
Resilience Factors			
Supportive family and friends	Supportive family and friends		Supportive family and friends
Support at work	Supervisor, colleagues		Mentor, collaborators, labmates
Feeling adequately trained	Trainings and supervision		Support to "COVID-ized" work
Structure in the workplace	Team approach		Structured office and home time
Safety in the workplace	Clear safety instructions		Clear safety instructions
Coping strategies	Humor and planning		Humor and planning
Confidants	Open dialog about stress		Open dialog about stress
Personal beliefs/meaning	Altruism and spirituality		Altruism and spirituality

NOTE: While the majority of research focusing on risk and resilience factors for posttraumatic stress and other mental health concerns has focused on health-care workers, many of the principals can be extrapolated to academic women in other STEMM fields.
SOURCE: Adapted from presentation by N. Epperson. *The Impact of COVID-19 on Mental Health and Wellbeing of Women in STEMM.* November 6, 2020.

altruism, ability to use humor, making plans, and being able to make meaning of the current situation are protective factors in the face of tremendous work-related stress (Carmassi et al., 2020).

CONCLUSION

The stress associated with the COVID-19 pandemic in 2020 accentuated the gender gap in STEMM. Unlike community-based stressors that have occurred in the United States since the last global pandemic in the early 20th century, the nature of the stressors related to the COVID-19 pandemic affected all nations, continued for months, and required social distancing to reduce viral spread. This type of chronic and unpredictable stress, along with social isolation, is particularly aversive for women in STEMM, worsening mental health concerns during the COVID-19 pandemic. Indeed, when it comes to women in health-care professions, particularly those providing care to SARS-CoV-2–infected patients, the risk for increased depression, anxiety, and sleep disturbance was even greater in 2020 than that for men in health care as well as the general population, which can likely be expanded to reflect conditions among women in academic STEMM.

The mental health disorders that arise de novo or are worsened during a widespread crisis such as the COVID-19 pandemic are more common among women and People of Color. COVID-19 pandemic-related factors conspire to increase these problems for women in academic STEMM. Social isolation, lack of or disconnection from women role models, previous and ongoing exposures to discrimination and related stress, biological and hormonal factors, and economic and family concerns are just a few of the larger social determinants of mental health among women in STEMM. Individual risk factors in risk for burnout, mental health concerns, and loss to the workforce are important to understand and consider.

8

Major Findings and Research Questions

INTRODUCTION

The COVID-19 pandemic, which began in late 2019, created unprecedented global disruption and infused a significant level of uncertainty into the lives of individuals, both personally and professionally, around the world throughout 2020. The significant effect on vulnerable populations, such as essential workers and the elderly, is well documented, as is the devastating effect the COVID-19 pandemic had on the economy, particularly brick-and-mortar retail and hospitality and food services. Concurrently, the deaths of unarmed Black people at the hands of law enforcement officers created a heightened awareness of the persistence of structural injustices in U.S. society.

Against the backdrop of this public health crisis, economic upheaval, and amplified social consciousness, an ad hoc committee was appointed to review the potential effects of the COVID-19 pandemic on women in academic science, technology, engineering, mathematics, and medicine (STEMM) during 2020. The committee's work built on the National Academies of Sciences, Engineering, and Medicine report *Promising Practices for Addressing the Underrepresentation of Women in Science, Engineering, and Medicine: Opening Doors* (the *Promising Practices* report), which presents evidence-based recommendations to address the well-established structural barriers that impede the advancement of women in STEMM. However, the committee recognized that none of the actions identified in the *Promising Practices* report were conceived within the context of a pandemic, an economic downturn, or the emergence of national protests against structural racism. The representation and vitality of academic women in STEMM had already warranted national attention prior to these events, and the COVID-19

109

pandemic appeared to represent an additional risk to the fragile progress that women had made in some STEMM disciplines. Furthermore, the future will almost certainly hold additional, unforeseen disruptions, which underscores the importance of the committee's work.

In times of stress, there is a risk that the divide will deepen between those who already have advantages and those who do not. In academia, senior and tenured academics are more likely to have an established reputation, a stable salary commitment, and power within the academic system. They are more likely, before the COVID-19 pandemic began, to have established professional networks, generated data that can be used to write papers, and achieved financial and job security. While those who have these advantages may benefit from a level of stability relative to others during stressful times, those who were previously systemically disadvantaged are more likely to experience additional strain and instability.

As this report has documented, during 2020 the COVID-19 pandemic had overall negative effects on women in academic STEMM in areas such productivity, boundary setting and boundary control, networking and community building, burnout rates, and mental well-being. The excessive expectations of caregiving that often fall on the shoulders of women cut across career timeline and rank (e.g., graduate student, postdoctoral scholar, non-tenure-track and other contingent faculty, tenure-track faculty), institution type, and scientific discipline. Although there have been opportunities for innovation and some potential shifts in expectations, increased caregiving demands associated with the COVID-19 pandemic in 2020, such as remote working, school closures, and childcare and eldercare, had disproportionately negative outcomes for women.

The effects of the COVID-19 pandemic on women in STEMM during 2020 are understood better through an intentionally intersectional lens. Productivity, career, boundary setting, mental well-being, and health are all influenced by the ways in which social identities are defined and cultivated within social and power structures. Race and ethnicity, sexual orientation, gender identity, academic career stage, appointment type, institution type, age, and disability status, among many other factors, can amplify or diminish the effects of the COVID-19 pandemic for a given person. For example, non-cisgender women may be forced to return to home environments where their gender identity is not accepted, increasing their stress and isolation, and decreasing their well-being. Women of Color had a higher likelihood of facing a COVID-19–related death in their family compared with their white, non-Hispanic colleagues. The full extent of the effects of the COVID-19 pandemic for women of various social identities was not fully understood at the end of 2020.

Considering the relative paucity of women in many STEMM fields prior to the COVID-19 pandemic, women are more likely to experience academic isolation, including limited access to mentors, sponsors, and role models that share gender, racial, or ethnic identities. Combining this reality with the physical isolation stipulated by public health responses to the COVID-19 pandemic, women in STEMM were subject to increasing isolation within their fields, networks, and communities. Explicit attention to the early indicators of how the COVID-19 pandemic affected women in academic STEMM careers during 2020, as well as attention to crisis responses throughout history, may provide opportunities to mitigate some of the long-term effects and potentially develop a more resilient and equitable academic STEMM system.

MAJOR FINDINGS

Given the ongoing nature of the COVID-19 pandemic, it was not possible to fully understand the entirety of the short- or long-term implications of this global disruption on the careers of women in academic STEMM. Having gathered preliminary data and evidence available in 2020, the committee found that significant changes to women's work-life boundaries and divisions of labor, careers, productivity, advancement, mentoring and networking relationships, and mental health and well-being have been observed. The following findings represent those aspects that the committee agreed have been substantiated by the preliminary data, evidence, and information gathered by the end of 2020. They are presented either as Established Research and Experiences from Previous Events or Impacts of the COVID-19 Pandemic during 2020 that parallel the topics as presented in the report.

Established Research and Experiences from Previous Events

Finding 1 **Women's Representation in STEMM:** Leading up to the COVID-19 pandemic, the representation of women has slowly increased in STEMM fields, from acquiring Ph.D.s to holding leadership positions, but with caveats to these limited steps of progress; for example, women representation in leadership positions tends to be at institutions with less prestige and fewer resources. While promising and encouraging, such progress is fragile and prone to setbacks especially in times of crisis (see Chapter 6).

Finding 2 **Confluence of Social Stressors:** Social crises (e.g., terrorist attacks, natural disasters, racialized violence, and infectious diseases) and COVID-19 pandemic-related disruptions to workload and schedules, added to formerly routine job functions and health risks, have the potential to exacerbate mental health conditions such as insomnia, depression, anxiety, and posttraumatic stress. All of these conditions occur more frequently among women than men.[1] As multiple crises coincided during 2020, there is a greater chance that women will be affected mentally and physically (see Chapters 4 and 7).

---

[1] This finding is primarily based on research on cisgender women and men.

Finding 3   **Intersectionality and Equity:** Structural racism is an omnipresent stressor for Women of Color, who already feel particularly isolated in many fields and disciplines. Attempts to ensure equity for all women may not necessarily create equity for women across various identities if targeted interventions designed to promote gender equity do not account for the racial and ethnic heterogeneity of women in STEMM (see Chapters 1, 3, and 4).

Impacts of the COVID-19 Pandemic during 2020

Finding 4   **Academic Productivity:** While some research indicates consistency in publications authored by women in specific STEMM disciplines, like Earth and space sciences, during 2020, several other preliminary measures of productivity suggest that COVID-19 disruptions have disproportionately affected women compared with men. Reduced productivity may be compounded by differences in the ways research is conducted, such as whether field research or face-to-face engagement with human subjects is required (see Chapter 3).

Finding 5   **Institutional Responses:** Many administrative decisions regarding institutional supports made during 2020, such as work-from-home provisions and extensions on evaluations or deliverables, are likely to exacerbate underlying gender-based inequalities in academic advancement rather than being gender neutral as assumed. For example, while colleges and universities have offered extensions for those on the tenure track and federal and private funders have offered extensions on funding and grants, these changes do not necessarily align with the needs expressed by women, such as the need for flexibility to contend with limited availability of caregiving and requests for a reduced workload, nor do they generally benefit women faculty who are not on the tenure track. Furthermore, provision of institutional support may be insufficient if it does not account for the challenges faced by those with multiple marginalized identities (see Chapters 3 and 4).

Finding 6   **Institutional Responses:** Organizational-level approaches may be needed to address challenges that have emerged as a result of the COVID-19 pandemic in 2020, as well as those challenges that may have existed before the pandemic but are now more visible and amplified. Reliance on individual coping strategies may be insufficient (see Chapters 2 and 6).

Finding 7   **Work-Life Boundaries and Gendered Divisions of Labor:** The COVID-19 pandemic has intensified complications related to work-life boundaries that largely affect women. Preliminary evidence from 2020 suggests women in academic STEMM are experiencing increased workload, decreased productivity, changes in interactions, and difficulties from remote work caused by the COVID-19 pandemic and associated disruptions. Combined with the gendered division of nonemployment labor that affected women before the pandemic, these challenges have been amplified, as demonstrated by a lack of access to childcare, children's heightened behavioral and academic needs, increased eldercare demands, and personal physical and mental health concerns. These are particularly salient for women who are parents or caregivers (see Chapter 4).

Finding 8 **Collaborations:** During the COVID-19 pandemic, technology has allowed for the continuation of information exchange and many collaborations. In some cases technology has facilitated the increased participation of women and underrepresented groups. However, preliminary indicators also show gendered impacts on science and scientific collaborations during 2020. These arise because some collaborations cannot be facilitated online and some collaborations face challenges including finding time in the day to engage synchronously, which presents a larger burden for women who manage the larger share of caregiving and other household duties, especially during the first several months of the COVID-19 pandemic (see Chapter 5).

Finding 9 **Networking and Professional Societies:** During the COVID-19 pandemic in 2020, some professional societies adapted to the needs of members as well as to broader interests of individuals engaged in the disciplines they serve. Transitioning conferences to virtual platforms has produced both positive outcomes, such as lower attendance costs and more open access to content, and negative outcomes, including over-flexibility (e.g., scheduling meetings at non-traditional work hours; last-minute changes) and opportunities for bias in virtual environments (see Chapter 5).

Finding 10 **Academic Leadership and Decision-Making:** During the COVID-19 pandemic in 2020, many of the decision-making processes, including financial decisions like lay-offs and furloughs, that were quickly implemented contributed to unilateral decisions that frequently deviated from effective practices in academic governance, such as those in crisis and equity-minded leadership. Fast decisions greatly affected contingent and nontenured faculty members— positions that are more often occupied by women and People of Color. In 2020, these financial decisions already had negative, short-term effects and may portend long-term consequences (see Chapter 6).

Finding 11 **Mental Health and Well-being:** Social support, which is particularly important during stressful situations, is jeopardized by the physical isolation and restricted social interactions that have been imposed during the COVID-19 pandemic. For women who are already isolated within their specific fields or disciplines, additional social isolation may be an important contributor to added stress (see Chapter 7).

Finding 12 **Mental Health and Well-being:** For women in the health professions, major risk factors during the COVID-19 pandemic in 2020 included unpredictability in clinical work, evolving clinical and leadership roles, the psychological demands of unremitting and stressful work, and heightened health risks to family and self (see Chapter 7).

RESEARCH QUESTIONS

While this report compiled much of the research, data, and evidence available in 2020 on the effects of the COVID-19 pandemic, future research is still needed to understand all the potential effects, especially any long-term implications. The research questions represent areas the committee identified for future research, rather than specific recommendations. They are presented in six categories that parallel the chapters of the report: Cross-Cutting Themes; Academic Productivity and Institutional Responses; Work-Life Boundaries and Gendered Divisions of Labor; Collaboration, Networking, and Professional Societies; Academic Leadership and Decision-Making; and Mental Health and Well-being. The committee hopes the report will be used as a basis for continued understanding of the impact of the COVID-19 pandemic in its entirety and as a reference for mitigating impacts of future disruptions that affect women in academic STEMM. The committee also hopes that these research questions may enable academic STEMM to emerge from the pandemic era a stronger, more equitable place for women. Therefore, the committee identifies two types of research questions in each category; listed first are those questions aimed at understanding the impacts of the disruptions from the COVID-19 pandemic, followed by those questions exploring the opportunities to help support the full participation of women in the future.

Cross-Cutting Themes

1. What are the short- and long-term effects of the COVID-19 pandemic on the career trajectories, job stability, and leadership roles of women, particularly of Black women and other Women of Color? How do these effects vary across institutional characteristics,[2] discipline, and career stage?

---

[2] Institutional characteristics include different institutional types (e.g., research university, liberal arts college, community college), locales (e.g., urban, rural), missions (e.g., Historically Black Colleges and Universities, Hispanic-Serving Institutions, Asian American/Native American/Pacific Islander-Serving Institutions, Tribal Colleges and Universities), and levels of resources.

MAJOR FINDINGS AND RESEARCH QUESTIONS

2. How did the confluence of structural racism, economic hardships, and environmental disruptions affect Women of Color during the COVID-19 pandemic? Specifically, how did the murder of George Floyd, Breonna Taylor, and other Black citizens impact Black women academics' safety, ability to be productive, and mental health?
3. How has the inclusion of women in leadership and other roles in the academy influenced the ability of institutions to respond to the confluence of major social crises during the COVID-19 pandemic?
4. How can institutions build on the involvement women had across STEMM disciplines during the COVID-19 pandemic to increase the participation of women in STEMM and/or elevate and support women in their current STEMM-related positions?
5. How can institutions adapt, leverage, and learn from approaches developed during 2020 to attend to challenges experienced by Women of Color in STEMM in the future?

Academic Productivity and Institutional Responses

6. How did the institutional responses (e.g., policies, practices) that were outlined in the Major Findings impact women faculty across institutional characteristics and disciplines?
7. What are the short- and long-term effects of faculty evaluation practices and extension policies implemented during the COVID-19 pandemic on the productivity and career trajectories of members of the academic STEMM workforce by gender?
8. What adaptations did women use during the transition to online and hybrid teaching modes? How did these techniques and adaptations vary as a function of career stage and institutional characteristics?
9. What are examples of institutional changes implemented in response to the COVID-19 pandemic that have the potential to reduce systemic barriers to participation and advancement that have historically been faced by academic women in STEMM, specifically Women of Color and other marginalized women in STEMM? How might positive institutional responses be leveraged to create a more resilient and responsive higher education ecosystem?
10. How can or should funding arrangements be altered (e.g., changes in funding for research and/or mentorship programs) to support new ways of interaction for women in STEMM during times of disruption, such as the COVID-19 pandemic?

Work-Life Boundaries and Gendered Divisions of Labor

11. How do different social identities (e.g., racial; socioeconomic status; culturally, ethnically, sexually, or gender diverse; immigration status; parents of young children and other caregivers; women without partners) influence the management of work-nonwork boundaries? How did this change during the COVID-19 pandemic?
12. How have COVID-19 pandemic-related disruptions affected progress toward reducing the gender gap in academic STEMM labor-force participation? How does this differ for Women of Color or women with caregiving responsibilities?
13. How can institutions account for the unique challenges of women faculty with parenthood and caregiving responsibilities when developing effective and equitable policies, practices, or programs?
14. How might insights gained about work-life boundaries during the COVID-19 pandemic inform how institutions develop and implement supportive resources (e.g., reductions in workload, on-site childcare, flexible working options)?

Collaboration, Networking, and Professional Societies

15. What were the short- and long-term effects of the COVID-19 pandemic-prompted switch from in-person conferences to virtual conferences on conference culture and climate, especially for women in STEMM?
16. How will the increase in virtual conferences specifically affect women's advancement and career trajectories? How will it affect women's collaborations?
17. How has the shift away from attending conferences and in-person networking changed longer-term mentoring and sponsoring relationships, particularly in terms of gender dynamics?
18. How can institutions maximize the benefits of digitization and the increased use of technology observed during the COVID-19 pandemic to continue supporting women, especially marginalized women, by increasing accessibility, collaborations, mentorship, and learning?
19. How can organizations that support, host, or facilitate online and virtual conferences and networking events (1) ensure open and fair access to participants who face different funding and time constraints; (2) foster virtual connections among peers, mentors, and sponsors; and (3) maintain an inclusive environment to scientists of all backgrounds?
20. What policies, practices, or programs can be developed to help women in STEMM maintain a sense of support, structure, and stability during and after periods of disruption?

Academic Leadership and Decision-Making

21. What specific interventions did colleges and universities initiate or prioritize to ensure that women were included in decision-making processes during responses to the COVID-19 pandemic?
22. How effective were colleges and universities that prioritized equity-minded leadership, shared leadership, and crisis leadership styles at mitigating emerging and potential negative effects of the COVID-19 pandemic on women in their communities?
23. What specific aspects of different leadership models translated to more effective strategies to advance women in STEMM, particularly during the COVID-19 pandemic?
24. How can examples of intentional inclusion of women in decision-making processes during the COVID-19 pandemic be leveraged to develop the engagement of women as leaders at all levels of academic institutions?
25. What are potential "top-down" structural changes in academia that can be implemented to mitigate the adverse effects of the COVID-19 pandemic or other disruptions?
26. How can academic leadership, at all levels, more effectively support the mental health needs of women in STEMM?

Mental Health and Well-being

27. What is the impact of the COVID-19 pandemic and institutional responses on the mental health and well-being of members of the academic STEMM workforce as a function of gender, race, and career stage?
28. How are tools and diagnostic tests to measure aspects of well-being, including burnout and insomnia, used in academic settings? How does this change during times of increased stress, such as the COVID-19 pandemic?
29. How might insights gained about mental health during the COVID-19 pandemic be used to inform preparedness for future disruptions?
30. How can programs that focus on changes in biomarkers of stress and mood dysregulation, such as levels of sleep, activity, and texting patterns, be developed and implemented to better engage women in addressing their mental health?
31. What are effective interventions to address the health of women academics in STEMM that specifically account for the effects of stress on women? What are effective interventions to mitigate the excessive levels of stress for Women of Color?

Glossary

Academic productivity Contributions to higher education and scholarship, typically organized into three categories:

Research, which includes scholarly productivity such as the generation of scientific papers, citations, grants, patents, and scholarly awards (Way et al., 2019).

Teaching, which includes creative activities associated with education and teaching; authoring textbooks; development of learning modules and tools; scheduled teaching obligations and course load; and informal tutoring, mentoring, and sponsorship of students and trainees.

Service, which includes internal service to one's department, school, or university in activities related to faculty governance, recruitment, student admissions, and program development; and external service to the profession and to local, state, national, or international communities, such as service on review committees, advisory boards, and editorial boards, as well as review of publications (Guarino and Borden, 2017).

Academic STEMM workforce For purposes of this report, all individuals in a STEMM field employed at a college or university in an academic position including tenure-track and non-tenure-track faculty, among others, with teaching, research, clinical, outreach, extension, or other "engagement" responsibilities. Where noted, this may also include postdoctoral researchers and graduate students.

Boundary control An employee's ability to manage the distinctions between work and nonwork roles. Boundary control includes employees' ability to maintain the boundary according to their personal and professional preferences (Kossek et al., 2012; Wotschack et al., 2014).

Burnout "A syndrome conceptualized as resulting from chronic workplace stress that has not been successfully managed. It is characterized by three dimensions: (1) feelings of energy depletion or exhaustion; (2) increased mental distance from one's job, or feelings of negativism or cynicism related to one's job; and (3) reduced professional efficacy. Burnout refers specifically to phenomena in the occupational context and should not be applied to describe experiences in other areas of life" (WHO, 2019).

Collaboration An act of engagement with individuals from one's network toward the pursuit of a shared or common research, teaching, or educational agenda, such as when a group of at least two scholars author academic papers together. Collaborators are a subset of one's professional network and, most often, are those with the closest shared interest to the individual researcher (Newman, 2000).

Contingent staff Outsourced and nonpermanent workers who are hired on a per-project basis. Salary may or may not be guaranteed depending on the state of the project.

Crisis leadership The process by which an entity (individual, organization, etc.) responds to a major unpredictable event that is threatening to cripple an organization, by employing practices that connect with people broadly as individuals and establishing mutual trust, distributing leadership throughout the organization, and communicating clearly and often with all stakeholders (Fernandez and Shaw, 2020).

Employee well-being	A characterization of quality of life with respect to the work-related environmental, organizational, and psychosocial factors that influence an individual's health. "Well-being is the experience of positive perceptions and the presence of constructive conditions at work and beyond that enables workers to thrive and achieve their full potential" (Chari et al., 2018, p. 590).
Equity	A solution for addressing imbalanced social systems that recognizes that each person has different circumstances and allocates the exact resources and opportunities needed to reach an equal outcome. The World Health Organization defines equity as "the absence of avoidable or remediable differences among groups of people, whether those groups are defined socially, economically, demographically or geographically" (WHO, 2021).
Equity-minded leadership	A leadership style characterized by being evidence based (i.e., using data), race conscious, institutionally focused, systemically aware, and equity advancing (Dowd and Bensimon, 2014).
Extensions	Additional time granted to complete expected tasks. For faculty, extensions might be provided during their probationary period (pre-tenure) to meet the demands of scholarship required by promotion and tenure policies or by a funding agency to meet the agreed-upon terms of a grant. For graduate students, extensions might be provided to allow for additional time to complete a milestone, such as degree completion. Finally, extensions can be awarded to accommodate special circumstances (e.g., childbirth or adoption).
Gender	A nonbinary social construct that refers to social and cultural differences rather than biological differences. Gender interacts with but is different from sex, which is a biological trait (Butler, 2004; Laner, 2000; Pichevin and Hurtig, 2007).
Gig Academy	A term to describe how corporate gig economy practices, such as hiring contingent labor or outsourcing, have been adopted by colleges and universities across the country (Kezar et al., 2019).

Institution type	Refers to both Carnegie Classifications of Institutions of Higher Education (e.g., doctoral universities, master's colleges and universities) and categories of Minority-Serving Institutions such as Historically Black Colleges and Universities, Hispanic-Serving Institutions, or Tribal Colleges and Universities.
Intersectionality	A field of study and an analytic lens that makes visible the mutually constructive and reciprocal relationship among race, ethnicity, sexuality, class, and other social positions that influence one's experiences (Collins, 2015). In this report, it is used as a lens to help understand how social identities, especially for marginalized groups, relate to systems of authority and power. Intersectionality is rooted in Black feminism and Critical Race Theory: in reference to historic exclusion of Black women, legal scholar Kimberlé Williams Crenshaw used intersectionality to describe the intersection of gender and race discrimination, arguing that treating them as exclusive, and not intertwined, renders the multiple marginalities faced by Black women invisible to antidiscrimination law (Carbado, 2013; Crenshaw, 1989, 1991, 2014).
Mentor	A person who provides both career and psychosocial support to help guide personal and professional growth over time.
Mentorship	A professional, working alliance in which individuals work together over time to support the personal and professional growth, development, and success of the relational partners through the provision of career and psychosocial support (NASEM, 2019c).
Networking	The act of growing one's connections and visibility within a field through formal and informal interactions with new colleagues who share similar research interests. Networking includes, but is not limited to, the act of engaging in professional organization conferences, academic seminars, and other related activities (Mickey, 2019).

People of Color An evolving term that emerged in the 1960s, which now includes a broader group of individuals such as Black, Latinx, Asian, Mexican, Japanese, Chinese, and other groups that share a common societal place of either feeling or actually being marginalized (Pérez, 2020). For purposes of this report, People of Color refers to all individuals specifically identifying as Black, Latinx, and American Indians/Alaska Natives.

Professional societies Organizations that facilitate networking and collaboration by providing a shared or common space for people in similar disciplines to interact, share information, and learn about each other's research through conferences and other society-sponsored activities. They may also be organized around nondisciplinary interests or demographic affinity (NAE, 2017). As such, these bodies play a pivotal role in facilitating mentorship and sponsorship programs for individuals within the same professional fields and serve as the primary source of postcollegiate education and skills training for the workforce (ASAE, 2020; Cree-Green et al., 2020). STEMM-based societies and organizations are critical to maintaining strong networks and systems of collaboration, as they provide an opportunity for academics and scientists with shared interests to interact, learn about each other's research, spark ideas for future collaborations, and lend assistance in moments of great need, such as after natural disasters (Cree-Green et al., 2020).

Sex Sex refers to the different biological and physiological characteristics of males, females, and intersex persons (Butler, 2004; Laner, 2000; Pichevin and Hurtig, 2007).

Shared leadership A leadership style characterized by four key elements (Kezar and Holcombe, 2017):
- A greater number of individuals take on leadership roles than in traditional models.
- Leaders and followers are seen as interchangeable. In some cases this may mean that leadership occurs on a flexible and emergent basis, while in others it rotates more formally.
- Leadership is not based on position or authority. Rather, individuals with the expertise and skills

needed for solving the problem at hand are those that lead. To that end, multiple perspectives and expertise are capitalized on for problem-solving, innovation, and change.
- Collaboration and interactions across the organization are typically emphasized.

Sponsor — Someone with power or in a leadership position who can use their resources (e.g., financial, professional), power, influence, or stature to advocate for the advancement or visibility of an individual (Catalyst, 2020b; Rockquemore, 2015).

Sponsorship — An act involving a senior person publicly acknowledging the achievements of and advocating for a mentee (Kram, 1985; Ragins and McFarlin, 1990).

Woman — Any person who identifies as a woman, including, but not limited to, cisgender women, transgender women, and nonbinary women. Woman refers to a person's gender, whereas female refers to a person's sex. Although these terms are often used interchangeably and can be related, they are discrete concepts.

Work-family enrichment — The positive transfer of knowledge, skills, and emotions from work experiences to family experiences or the reverse (Greenhaus and Powell, 2006).

Work-life boundary management — The control held between work and nonwork roles that can be reinforced or weakened cognitively, physically, and emotionally (Allen et al., 2014; Ashforth et al., 2000; Kossek et al., 2012).

Work-life integration — Also referred to as "work-life balance," the incorporation of work and nonwork demands. This involves subjective assessments of balance and satisfaction in both work and nonwork roles (e.g., Greenhaus and Allen, 2011) and structural or organizational factors that affect this integration or balance, such as flexibility in work schedules and tenure assessments (e.g., Kossek and Lambert, 2005; Moss et al., 2005).

Workplace diversity — The demographic variation within an organization's staff and leadership.

References

AAMC (Association of American Medical Colleges) (2016). Diversity in medical education. *AAMC Facts and Figures*. https://www.aamcdiversityfactsandfigures2016.org.

AAMC (2020). Affinity Groups. https://www.aamc.org/professional-development/affinity-groups.

AAUP (American Association of University Professors) (2007). *Report of an AAUP Special Committee: Hurricane Katrina and New Orleans Universities*. Academe: AAUP Bulletin: 60-126. https://www.aaup.org/report/report-aaup-special-committee-hurricane-katrina-and-new-orleans-universities.

AAUP (2017). Data visualization of faculty by contract type and institutional type. https://www.aaup.org/data-visualizations-contingent-faculty-us-higher-education.

AAUP (2018). The Annual Report on the Economic Status of the Profession, 2017-18. https://www.aaup.org/sites/default/files/ARES_2017-18.pdf.

AAUP (2020a). Annual Report on the Economic Status of the Profession, 2019–2020. https://www.aaup.org/sites/default/files/2019-20_ARES.pdf.

AAUP (2020b). Full-Time Women Faculty and Faculty of Color, December 9, 2020. https://www.aaup.org/news/data-snapshot-full-time-women-faculty-and-faculty-color.

ACE (American Council on Education) (2017). *Pipelines, Pathways, and Institutional Leadership: An Update on the Status of Women in Higher Education*. http://www.acenet.edu/news-room/Documents/HES-Pipelines-Pathways-and-Institutional-Leadership-2017.pdf.

Adams, S. (2020). ASEE Annual Conference: Financial Support Available (email to membership). May 1, 2020.

Ahlers-Schmidt, C. R., Hervey, A. M., Neil, T., Kuhlmann, S., and Kuhlmann, Z. (2020). Concerns of women regarding pregnancy and childbirth during the COVID-19 pandemic. *Patient Education and Counseling*. https://doi.org/10.1016/j.pec.2020.09.031.

Ahmed, S. (2012). *On Being Included: Racism and Diversity in Institutional Life*. Durham, NC: Duke University Press.

Akkermans, J., Richardson, J., and Kraimer, M. (2020). The COVID-19 crisis as a career shock: Implications for careers and vocational behavior. *Journal of Vocational Behavior, 119*. https://doi.org/10.1016/j.jvb.2020.103434.

Alexander, B. (2020). How the coronavirus will change faculty life forever: As the pandemic wears on, expect heavier teaching loads, more service requirements, and more time online. *Chronicle of Higher Education,* May 11, 2020. https://www.chronicle.com/article/how-the-coronavirus-will-change-faculty-life-forever/.

Allan, E. J. (2011). *Women's Status in Higher Education: Equity Matters* [Special Issue]. *ASHE (Association for the Study of Higher Education) Higher Education Report, 37*(1), 1–163.

Allegretto, S. A., and Cooper, D. (2014). Twenty-Three Years and Still Waiting for Change: Why It's Time to Give Tipped Workers the Regular Minimum Wage. Briefing Paper No. 379, July 10, 2014. Washington, DC: Economic Policy Institute. http://www.epi.org/publication/waiting-for-change-tipped-minimumwage/.

Allen, T. D., Cho, E., & Meier, L. L. (2014). Work–family boundary dynamics. *Annual Review of Organizational Psychology and Organizational Behavior*, 1, 99–121. https://doi.org/10.1146/annurev-orgpsych-031413-091330

Al-Omoush, K. S., Simón-Moya, V., and Sendra-García, J. (2020). The impact of social capital and collaborative knowledge creation on e-business proactiveness and organizational agility in responding to the COVID-19 crisis. *Journal of Innovation & Knowledge*, Vol 5, Iss 4, Pages 279-288. https://doi.org/10.1016/j.jik.2020.10.002.

Alon, T., Doepke, M., Olmstead-Rumsey, J., and Tertilt, M. (2020a). The Impact of COVID-19 on Gender Equality (NBER Working Paper Series No. 26947). Cambridge, MA: National Bureau of Economic Research.

Alon, T., Doepke, M., Olmstead-Rumsey, J., and Tertilt, M. (2020b). This Time It's Different: The Role of Women's Employment in a Pandemic Recession. Working Paper Series No. 27660. Cambridge, MA: National Bureau of Economic Research.

Amano-Patino, N., Faraglia, E., Giannitsarou, C., and Hasna, Z. (2020). Who is doing new research in the time of COVID-19? Not the female economists. *VoxEU,* May 2, 2020. https://voxeu.org/article/who-doing-new-research-time-COVID-19-not-female-economists.

Andersen, J. P., Nielsen, M. W., Simone, N. L., Lewiss, R. E., and Jagsi, R. (2020). Meta-Research: COVID-19 medical papers have fewer women first authors than expected. *eLife, Sciences*, e58807. https://doi.org/10.7554/eLife.58807.

Antecol, H., Bedard, K., and Stearns, J. (2018). Equal but inequitable: Who benefits from gender-neutral tenure clock stopping policies? *American Economic Review*, *108*(9), 2420–2441. https://doi.org/10.1257/aer.20160613.

Anwer, M. (2020). Academic labor and the global pandemic: Revisiting life-work balance under COVID-19. Working Paper Series, *Navigating Careers in the Academy: Gender, Race, and Class*, *3*(2), 5–13. https://www.purdue.edu/butler/documents/WPS-Special-Issue-Higher-Education-and-COVID-19—2020-Volume-3-Issue-2.pdf#page=8.

APM (American Public Media) Research Lab Staff (2020). The color of coronavirus: COVID-19 deaths by race and ethnicity in the U.S. Updated regularly. https://www.apmresearchlab.org/covid/deaths-by-race.

Apuzzo, M., and Kirkpatrick, D. D. (2020). "COVID-19 Changed How the World Does Science, Together." *New York Times*, April 14, 2020. https://nyti.ms/2Uy7HRD.

ARC (ADVANCE Research and Coordination [ARC] Network) (2020). Ensuring Equity in COVID-19 Institutional Responses (webinar notes and transcripts).

Armstrong, M. A., and Jovanovic, J. (2015). Starting at the crossroads: Intersectional approaches to institutionally supporting underrepresented minority women STEM faculty. *Journal of Women and Minorities in Science and Engineering*, *21*(2), 141–157.

Armstrong, M. A., and Jovanovic, J. (2017). The intersectional matrix: Rethinking institutional change for URM women in STEM. *Journal of Diversity in Higher Education*, *10*(3), 216–231.

ASAE (American Society of Association Executives) (2020). *Association Impact Snapshot March/April 2020: Trending Report.* Washington, DC: ASAE Research Foundation.

REFERENCES

ASCCC (Academic Senate for California Community Colleges). (2010). *Practices that Promote Equity in Basic Skills in California Community Colleges*. https://www.asccc.org/sites/default/files/publications/promote_equity_basicskills-spr2010_0.pdf

ASEE (American Society for Engineering Education) (2020). *COVID-19 & Engineering Education: An Interim Report on the Community Response to the Pandemic and Racial Justice*. Washington, DC: American Society for Engineering Education.

Ashforth, B. E., Kreiner, G. E., and Fugate, M. (2000). All in a day's work: Boundaries and micro role transitions. *Academy of Management Review*, 25, 472–491. https://doi.org/10.5465/amr.2000.3363315.

ASM (American Society for Microbiology) (2020). Press Release. ASM COVID-19 Initiatives. https://asm.org/Press-Releases/2020/ASM-Initiatives-for-COVID-19.

Aspire, The National Alliance for Inclusive and Diverse STEM Faculty (2020). COVID-19 Resources. https://www.aspirealliance.org/national-change/COVID-19-resources.

Attridge, M. (2007). Making the business case. Plenty of studies prove that employers should invest in their workers' mental well-being. *Behavioral Healthcare*, 27(11):31–33. PMID: 18293789.

Attridge, M. (2009). Measuring and managing employee work engagement: A review of the research and business literature. *Journal of Workplace Behavioral Health*, 24, 383–398. https://doi.org/10.1080/15555240903188398.

Aubry, L., Laverty, T., and Ma, Z. (2020). Impacts of COVID-19 on ecology and evolutionary biology faculty in the United States. *Ecological Applications*, November 23, 2020. https://esajournals.onlinelibrary.wiley.com/doi/10.1002/eap.2265.

Avery-Gomm, S., Hammer, S., and Humphries, G. (2016). The age of the Twitter conference. *Science* 352(6292), 1404–1405. https://doi.org/10.1126/science.352.6292.1404-b.

Aviles, G. (2020). "The coronavirus is threatening diversity in academia." *NBC News*, May 25, 2020. https://www.nbcnews.com/news/us-news/coronavirus-threatening-diversity-academia-n1212931.

Bacher-Hicks, A., Goodman, J., and Mulhern, C. (2020). Inequality in Household Adaptation to Schooling Shocks: Covid-Induced Online Learning Engagement in Real Time. NBER Working Paper No. 27555, July 2020, revised November 2020. JEL No. I20,I24. Cambridge, MA: National Bureau of Economic Research.

Banasik, M. D., and Dean, J. (2015). Non-tenure track faculty and learning communities: Bridging the divide to enhance teaching quality. *Innovative Higher Education*, 41(4), 333–342. https://doi.org/10.1007/s10755-015-9351-6.

Barak, A. (2005). Sexual harassment on the internet. *Social Science Computer Review*, 23(1), 77–92. https://journals.sagepub.com/doi/pdf/10.1177/0894439304271540.

Bauerlein, M., Gad-el-Hak, M., Grody, W., McKelvey, B., and Trimble, S. W. (2010). We must stop the avalanche of low-quality research, *Chronicle of Higher Education*, June 13, 2010.

Bauman, D. (2020). The pandemic has pushed hundreds of thousands of workers out of higher education. *Chronicle of Higher Education*, October 6, 2020. https://www.chronicle.com/article/how-the-pandemic-has-shrunk-higher-educations-work-force.

Beede, D. N., Julian, T. A., Langdon, D., McKittrick, G., Khan, B., and Doms, M. E. (2011). Women in STEM: A gender gap to innovation. SSRN Economics and Statistics Administration Issue Brief No. 04-11, November 26, 2011.

Belay, K. (2020). What has higher education promised on anti-racism in 2020 and is it enough? EAB, November 2020. https://eab.com/research/expert-insight/strategy/higher-education-promise-anti-racism/.

Bell, E. (2020). How Associations Are Using Community to Support Their Members during COVID-19. Arlington, VA: Higher Logic. https://www.higherlogic.com/blog/associations-community-support-members-COVID-19/.

Bellas, M. L. (1999). Emotional labor in academia: The case of professors. *Annals of the American Academy of Political and Social Science*, 561(1), 96–110.

Benchekroun, S., and Kuepper, M. (2020). "Guest Post – Coronavirus is a Wakeup Call for Academic Conferences. Here's Why." *The Scholarly Kitchen* (blog), posted March 25, 2020. https://scholarlykitchen.sspnet.org/2020/03/25/guest-post-coronavirus-is-a-wakeup-call-for-academic-conferences-heres-why/.

Berthelot, N., Lemieux, R., Garon-Bissonnette, J., Drouin-Maziade, C., Martel, E., and Maziade, M. (2020). Uptrend in distress and psychiatric symptomatology in pregnant women during the coronavirus disease 2019 pandemic. *Acta Obstetrica et Gynecologica Scandinavica*, 99(7), 848–855. https://doi.org/10.1111/aogs.13925.

Bérubé, M., and Ruth, J. (2015). *The Humanities, Higher Education, and Academic Freedom: Three Necessary Arguments*. London, UK: Palgrave Macmillan. https://doi.org/10.1057/9781137506122.

Bianchi, S. M., Sayer, L. C., Milkie, M. A., and Robinson, J. P. (2012). Housework: Who did, does or will do it, and how much does it matter? *Social Forces*, 91(1), 55–63. https://doi.org/10.1093/sf/sos120.

Bickel, J. 2004. Gender equity in undergraduate medical education: A status report. *Journal of Women's Health & Gender-Based Medicine*, 10(3), 261–270.

Bilen-Green, C., and Froelich, K. A. (2010). Women in university leadership positions: Comparing institutions with strong and weak gender equity track records. Paper presented at Women in Engineering ProActive Network.

Bilen-Green, C., Froelich, K. A., and Jacobson, S. W. (2008). "The prevalence of women in academic leadership positions, and potential impact on prevalence of women in the professorial ranks." 2008 Women in Engineering ProActive Network Conference Proceedings. https://www.ndsu.edu/fileadmin/forward/documents/WEPAN2.pdf.

Bilimoria, D., and Liang, X. (2012). *Gender Equity in Science and Engineering: Advancing Change in Higher Education*. Milton Park, Oxfordshire, UK: Routledge.

Bixler, D. (2020). SARS-CoV-2–Associated Deaths Among Persons Aged 21 Years—United States, February 12–July 31, 2020. MMWR: Morbidity and Mortality Weekly Report, 69.

Black in Engineering (2020). *On Becoming an Anti-Racist Institution*. https://blackinengineering.org/action-item-list/.

Blair-Loy, M., and Cech, E. A. (2017). Demands and devotion: Cultural meanings of work and overload among women researchers and professionals in science and technology industries. *Social Forces*, 32, 5–27. https://doi.org/10.1111/socf.12315.

Blevins, C. A., Weathers, F. W., Davis, M. T., Witte, T. K., and Domino, J. L. (2015). The post-traumatic stress disorder checklist for DSM-5 (PCL-5): Development and initial psychometric evaluation. *Journal of Traumatic Stress*, 28(6), 489–498. https://doi.org/10.1002/jts.22059. Epub 2015 Nov 25. PMID: 26606250.

Boland, W., and Gasman, M. (2014). *America's Public HBCUs: A Four State Comparison of Institutional Capacity and State Funding Priorities*. Penn Center for Minority Serving Institutions. http://repository.upenn.edu/gse_pubs/340.

Boserup, B., McKenney, M., and Elkbuli, A. (2020). Alarming trends in US domestic violence during the COVID-19 pandemic. *American Journal of Emergency Medicine*, 38(12), 2753–2755. https://doi.org/10.1016/j.ajem.2020.04.077.

Boswell, W. R., Olson-Buchanan, J. B., Butts, M. M., and Becker, W. J. (2016). Managing "after hours" electronic work communication. *Organizational Dynamics*, 45, 291–297. https://doi.org/10.1016/j.orgdyn.2016.10.004.

Bovin, M. J., Marx, B. P., Weathers, F. W., Gallagher, M. W., Rodriguez, P., Schnurr, P. P., and Keane, T. M. (2016). Psychometric properties of the PTSD Checklist for Diagnostic and Statistical Manual of Mental Disorders–Fifth Edition (PCL-5) in veterans. *Psychological Assessment*, 28(11), 1379–1391. https://doi.org/10.1037/pas0000254. Epub 2015 Dec 14. PMID: 26653052.

Bracke, P., Delaruelle, K., Dereuddre, R., and Van de Velde, S. (2020). Depression in women and men, cumulative disadvantage and gender inequality in 29 European countries. *Social Science & Medicine*, 113354. https://doi.org/10.1016/j.socscimed.2020.113354. Epub ahead of print. PMID: 32980172.

REFERENCES

Breslau, J., Cefalu, M., Wong, E. C., Burnam, M. A., Hunter, G. P., Florez, K. R., and Collins, R. L. (2017). Racial/ethnic differences in perception of need for mental health treatment in a US national sample. *Social Psychiatry and Psychiatric Epidemiology*, 52, 1–9.

Bridges, B. K., Eckel, P. D., Cordova, D. I., and White, B. P. (2007). *Broadening the Leadership Spectrum: Advancing Diversity in the American College Presidency*. American Council on Education.

Brockmeier, E. K. (2020). Research returns to campus. *Penn Today*, June 8, 2020, University of Pennsylvania. https://penntoday.upenn.edu/news/phase-i-research-resumption.

Brookings Institution (2017). Black women are earning more college degrees, but that alone won't close race gaps. *Social Mobility Memos* (blog), posted December 4, 2017. https://www.brookings.edu/blog/social-mobility-memos/2017/12/04/black-women-are-earning-more-college-degrees-but-that-alone-wont-close-race-gaps/.

Brooks, S. (2020). "Housing challenges and Black faculty." *Inside Higher Education*, July 16, 2020. https://www.insidehighered.com/views/2020/07/16/black-tenure-track-faculty-members-face-challenges-looking-housing-near-their.

Brubaker, L. (2020). Women Physicians and the COVID-19 Pandemic. *Journal of the American Medical Association*, 324(9), 835–836. https://doi.org/10.1001/jama.2020.14797

Buckee, C., Hedt-Gauthier, B., Mahmud, A., Martinez, P., Tedijanto, C., Murray, M., Khan, R., Menkir, T., Li, R., Suliman, S., Fosdick, B. K., Cobey, S., Rasmussen, A., Popescu, S., Cevik, M., Dada, S., Jenkins, H., Clapham, H., Mordecai, E., Hampson, K., Majumder, M. S., Wesolowski, A., Kuppalli, K., Rodriguez Barraquer, I., Smith, T. C., Hodcroft, E. B., Christofferson, Re. C., Gerardin, J., Eggo, R., Cowley, L., Childs, L. M., Keegan, L. T., Pitzer, V., Oldenburg, C., and Dhatt, R. (2020). Women in science are battling both COVID-19 and the patriarchy. *Times Higher Education*, May15, 2020. https://www.timeshighereducation.com/blog/women-science-are-battling-both-COVID-19-and-patriarchy.

Butler, J. (2004). *Undoing Gender*. New York: Routledge.

Buysse, D. J., Reynolds, C. F. III, Monk, T. H., Berman, S. R., and Kupfer, D. J. (1989). The Pittsburgh Sleep Quality Index: A new instrument for psychiatric practice and research. *Psychiatry Research*, 2(May 28), 193–213. https://doi.org/10.1016/0165-1781(89)90047-4. PMID: 2748771.

Cabarkapa, S., Nadjidai, S. E., Murgier, J., and Ng, C. H. (2020). The psychological impact of COVID-19 and other viral epidemics on frontline healthcare workers and ways to address it: A rapid systematic review. *Brain Behavior & Immunity – Health*, 8(October 2020), 100144. https://doi.org/10.1016/j.bbih.2020.100144.

Calisi, R. M., and a Working Group of Mothers in Science (2018). Opinion: How to tackle the childcare–conference conundrum. *Proceedings of the National Academy of Sciences of the United States of America*, 115(12), 2845–2849; first published March 5, 2018. https://doi.org/10.1073/pnas.1803153115.

Carbado, D. W. (2013). Colorblind intersectionality. *Signs: Journal of Women in Culture and Society*, 38(4), 811–845.

Cardel, M. I., Dean, N., and Montoya-Williams, D. (2020a). Preventing a secondary epidemic of lost early career scientists: Effects of COVID-19 pandemic on women with children. *Annals of the American Thoracic Society*, 17(11). https://doi.org/10.1513/AnnalsATS.202006-589IP.

Cardel, M. I., Dhurandhar, E., Yarar-Fisher, C., Foster, M., Hidalgo, B., McClure, L. A., Pagoto, S., Brown, N., Pekmezi, D., Sharafeldin, N., Willig A. L., and Angelini, C. (2020b). Turning chutes into ladders for women faculty: A review and roadmap for equity in academia. *Journal of Women's Health*, 29, 721–733.

Carlson, D., Petts, R., and Pepin, J. (2020a). Changes in parents' domestic labor during the COVID-19 pandemic. SocArXiv Papers. https://osf.io/preprints/socarxiv/jy8fn/.

Carlson, D., Petts, R., and Pepin, J. (2020b). "Men and Women Agree: During the COVID-19 Pandemic Men Are Doing More at Home." *Brief Reports, CCF News*, Council on Contemporary Families, posted on May 20, 2020. https://contemporaryfamilies.org/covid-couples-division-of-labor/.

Carmassi, C., Foghi, C., Dell'Oste, V., Cordone, A., Bertelloni, C. A., Bui, E., and Dell'Osso, L. (2020). PTSD symptoms in healthcare workers facing the three coronavirus outbreaks: What can we expect after the COVID-19 pandemic. *Psychiatry Research*, 292, 113312.

Carr, R. M. (2020). Reflections of a Black woman physician-scientist. *Journal of Clinical Investigation*, 130(11).

Cassedy, J. G. (1997). African Americans and the American Labor Movement. *Federal Records and African American History*, 29(2), Summer 1997.

Castilla, E. J., and Benard, S. (2010). The paradox of meritocracy in organizations. *Administrative Science Quarterly*, 55, 543–576.

Catalyst (2020a). The impact of COVID-19 on workplace inclusion: Survey. July 15, 2020. https://www.catalyst.org/research/workplace-inclusion-COVID-19.

Catalyst (2020b). Sponsorship and mentoring: Ask Catalyst Express. https://www.catalyst.org/research/sponsorship-mentoring-resources/.

CDC (Centers for Disease Control and Prevention) (2020a). COVID-19 Hospitalization and Death by Race/Ethnicity. https://www.cdc.gov/coronavirus/2019-ncov/covid-data/investigations-discovery/hospitalization-death-by-race-ethnicity.html.

CDC (2020b). Corona Virus (COVID-19). https://www.cdc.gov/coronavirus/2019-ncov/index.html.

CDC (2020c). Considerations for Events and Gatherings. https://www.cdc.gov/coronavirus/2019-ncov/community/large-events/considerations-for-events-gatherings.html.

CDE (Colorado Department of Education) (2010). *Equity Toolkit for Administrators*. https://www.cde.state.co.us/postsecondary/equitytoolkit.

Cech, E. A., and Blair-Loy, M. (2019). The changing career trajectories of new parents in STEM. *Proceedings of the National Academy of Sciences of the United States of America*, 116(10), 4182–4187.

Chai, S., and Freeman, R. B. (2019). Temporary colocation and collaborative discovery: Who confers at conferences. *Strategic Management Journal*, 40, 2138–2164. https://doi.org/10.1002/smj.3062.

Chari, R., Chang, C. C., Sauter, S. L., Petrun Sayers, E. L., Cerully, J. L., Schulte, P., Schill, A. L., and Uscher-Pines, L. (2018). Expanding the paradigm of occupational safety and health: A new framework for worker well-being. *Journal of Occupational and Environmental Medicine*, 60(7):589–593.

Chellaraj, G., Maskus, K. E., and Mattoo, A. (2005). The Contribution of Skilled Immigration and International Graduate Students to U.S. Innovation. Policy Research Working Paper No. 3588. Washington, DC: World Bank. https://doi.org/10/1596/1813-9450-3588.

Cheryan, S., Davies, P. G., Plaut, V. C., and Steele, C. M. (2009). Ambient belonging: How stereotypical cues impact gender participation in computer science. *Journal of Personality & Social Psychology*, 97(6), 1045–1060. https://doi.org/10.1037/a0016239.

Chesley, N. (2005). Blurring boundaries? Linking technology use, spillover, individual distress, and family satisfaction. *Journal of Marriage and Family*, 67(5), 1237–1248. https://doi.org/10.1111/j.1741-3737.2005.00213.x.

Chronicle Staff. (2020). As COVID-19 pummels budgets, colleges are resorting to layoffs and furloughs. *Chronicle*, July 2, 2020. https://www.chronicle.com/article/were-tracking-employees-laid-off-or-furloughed-by-colleges.

Clark, D., Mickey, E. L., and Misra, J. (2020). Reflections on institutional equity for faculty in response to COVID-19. *Navigating Careers in the Academy: Gender, Race, and Class*, 92. [Special Issue]. Susan Bulkeley Butler Center for Leadership Excellence and ADVANCE. https://www.purdue.edu/butler/working-paper-series/2020/special%20issue.html.

Clark, T. R. (2020). "8 Ways to Manage Your Team While Social Distancing." *Harvard Business Review*, March 24, 2020. https://hbr.org/2020/03/8-ways-to-manage-your-team-while-social-distancing.

Cnaan, A., Laird, N. M., and Slasor, P. (1997). Using the general linear mixed model to analyse unbalanced repeated measures and longitudinal data. *Statistics in Medicine*, 16(20), 2349–2380. https://doi.org/10.1002/(SICI)1097-0258(19971030)16:20<2349::AID-SIM667>3.0.CO;2-E.

REFERENCES

Cohen, J. (2005). "Hurricane Katrina Update." Johns Hopkins Medicine, September 2, 2005. https://www.hopkinsmedicine.org/mediaII/katrina/aamc.html.

Collins, C., Landivar, L. C., Ruppanner, L., and Scarborough, W. J. (2020). COVID-19 and the gender gap in work hours. *Gender, Work, and Organization*, 28(S1):101–112. https://doi.org/10.1111/gwao.12506.

Collins, P. H. (2015). Intersectionality's definitional dilemmas. *Annual Review of Sociology*, 41(1), 1–20. https://doi.org/10.1146/annurev-soc-073014-112142.

Community Brands (2020). *From Disruption to Opportunity Industry Study: Association Trends 2020: What Members Want and How Virtual Engagement Is Creating New Paths for Growth*. https://go.communitybrands.com/rs/402-UVF-672/images/RST-CB-2020-10-DisruptionTo Opportunity-Study.pdf.

Connolly, J. (2020). "We need to rethink what counts for tenure now." *Inside Higher Education*, April 9, 2020. https://www.insidedhighered.com/advice/2020/04/09/COVID-19-demands-reconsideration-tenure-requirements-going-forward-opinion.

Connor, J., Madhavan, S., Mokashi, M., Amanuel, H., Johnson, N. R., Pace, L. E., and Bartz, D. (2020). Health risks and outcomes that disproportionately affect women during the COVID-19 pandemic: A review. *Social Science & Medicine*, 266(September 13, 2020), 113364. https://doi.org/10.1016/j.socscimed.2020.113364. Epub ahead of print.

Coury, S., Huang, J., Kumar, A., Prince, S., Krivokovich, A., and Yee, L. (2020). Women in the Workplace 2020: Corporate America is at a critical crossroads. McKinsey & Company. https://www.mckinsey.com/featured-insights/diversity-and-inclusion/women-in-the-workplace.

Craig, L., and Churchill, B. (2020). Dual-earner parent couples' work and care during COVID-19. *Gender, Work, and Organization*, 28(S1), 66–79. https://doi.org/10.1111/gwao.12497.

Creary, S. J. (2020). A Case for Investigating Gender and Work-Life Inclusion Among Black Women Faculty in Business Schools. In E. Kossek and K.-H. Lee (eds.), *Fostering Gender and Work-Life Inclusion for Faculty in Understudied Contexts: An Organizational Science Lens*, 41–45. West Lafayette, IN: Purdue e-Pubs. https://doi.org/10.5703/1288284317257. https://docs.lib.purdue.edu/worklifeinclusion/2018/idgwli/ 2/.

Cree-Green, M., Carreau, A.-M., Davis, S. M., Frohnert, B. I., Kaar, J. L., Ma, N. S., Nokoff, N. J., Reusch, J. E. B., Simon, S. L., and Nadeau, K. J. (2020). Peer mentoring for professional and personal growth in academic medicine. *Journal of Investigative Medicine*. Epub ahead of print (November 21, 2020), 1–7. https://doi.org/10.1136/jim-2020-001291.

Crenshaw, K. (1989). Demarginalizing the intersection of race and sex: A Black feminist critique of antidiscrimination doctrine, feminist theory and antiracist politics. *University of Chicago Legal Forum*, Vol. 1989, Article 8. https://chicagounbound.uchicago.edu/uclf/vol1989/iss1/8.

Crenshaw, K. W. (1991). Mapping the margins: Intersectionality, identity politics, and violence against women of color. *Stanford Law Review*, 43(6), 1241–1299.

Crenshaw, K. W. (2014). The structural and political dimensions of intersectional oppression. *Intersectionality: A Foundations and Frontiers Reader*, 17–22.

Cui, R., Ding, H., and Zhu, F. (2020). Gender inequality in research productivity during the COVID-19 pandemic. *Working Knowledge*, Harvard Business School. https://hbswk.hbs.edu/item/gender-inequality-in-research-productivity-during-the-COVID-19-pandemic.

Curtis, J. W. (2019). *The Annual Report on the Economic Status of the Profession, 2018–19*. Washington, DC: American Association of University Professors.

CUWFA (College and University Work-Life-Family Association) (n.d.). Mission. https://www.cuwfa.org/mission.

Dean, D. R., Bracken, S. J., and Allen, J. K. (2009). *Women in Academic Leadership: Professional Strategies, Personal Choices*. Sterling, VA: Stylus Publishing.

DeCoux, M. (2005). Acute versus primary care: The health care decision making process for individuals with severe mental illness. *Issues in Mental Health Nursing*, 26(9), 935–951.

Del Boca, D., Oggero, N., Profeta, P., and Rossi, M. (2020). Women's and men's work, housework and childcare, before and during COVID-19. *Review of Economics of the Household 18*(4), 1001–1017. https://doi.org/10.1007/s11150-020-09502-1.

Desai, M. K., and Brinton, R. D. (2019). Autoimmune disease in women: Endocrine transition and risk across the lifespan. *Frontiers in Endocrinology, 10*(April 29), 265.

Dettmers, J. (2017). How extended work availability affects well-being: The mediating roles of psychological detachment and work-family-conflict. *Work & Stress, 31*(1), 24–41. https://doi.org/10.1080/02678373.2017.1298164.

Dettmers, J., Vahle-Hinz, T., Bamberg, E., Friedrich, N., and Keller, M. (2016). Extended work availability and its relation with start-of-day mood and cortisol. *Journal of Occupational Health Psychology, 21*(1), 105–118. https://doi.org/10.1037/a0039602.

Devi, S. (2020). Travel restrictions hampering COVID-19 response. *Lancet 395*(10233), 1331–1332. https://doi.org/10.1016/S0140-6736(20)30967-3.

Dong, L., Agnew, J., Mojtabai, R., Surkan, P. J., and Spira, A. P. (2017). Insomnia as a predictor of job exit among middle-aged and older adults: Results from the Health and Retirement Study. *Journal of Epidemiology and Community Health, 71*(8), 750–757, August 2017. https://doi.org/10.1136/jech-2016-208630. Epub March 15, 2017. PMID: 28298414.

Dowd, A. C., and Bensimon, E. M. (2014). *Engaging the Race Question: Accountability and Equity in U.S. Higher Education.* New York: Teachers College Press.

Dumas, T. L., and Sanchez-Burks, J. (2015). The professional, the personal, and the ideal worker: Pressures and objectives shaping the boundary between life domains. *Academy of Management Annals, 9*(1), 803–843. https://doi.org/10.1080/19416520.2015.1028810.

Dumas, T. L., Phillips, K. W., and Rothbard, N. P. (2013). Getting closer at the company party: Integration experiences, racial dissimilarity, and workplace relationships. *Organization Science, 24*(5), 1377–1401. https://doi.org/10.1287/orsc.1120.0808.

Duxbury, L., Higgins, C., Smart, R., and Stevenson, M. (2014). Mobile technology and boundary permeability: Mobile technology and boundary permeability. *British Journal of Management, 25*(3), 570–588. https://doi.org/10.1111/1467-8551.12027.

Eddleston, K. A., Mulki, J., and Clair, J. (2017). Toward understanding remote workers' management of work-family boundaries: The complexity of workplace embeddedness. *Group & Organization Management, 42*(3), 346–387. https://doi.org/10.1177/1059601115619548.

Eddy, P. L., Ward, K., and Khwaja, T. (eds.) (2017). *Critical Approaches to Women and Gender in Higher Education.* New York: Springer Nature.

El-Alayli, A., Hansen-Brown, A. A., and Ceynar, M. (2018). Dancing backwards in high heels: Female professors experience more work demands and special favor requests, particularly from academically entitled students. *Sex Roles, 79*(3-4), 136–150. https://doi.org/10.1007/s11199-017-0872-6.

Entmacher, J., Frohlich, L., Robbins, K., Martin, E., and Watson, L. (2014). *Underpaid & Overloaded: Women in Low-Wage Jobs.* National Women's Law Center. http://www.nwlc.org/resource/underpaid-overloaded-women-low-wage-jobs.

Ertl, B., Luttenberger, S., and Paechter, M. (2017). The impact of gender stereotypes on the self-concept of female students in STEM subjects with an under-representation of females. *Frontiers in Psychology, 8*, 703.

Esterwood, E., and Saeed, S. A. (2020). Past epidemics, natural disasters, COVID-19, and mental health: Learning from history as we deal with the present and prepare for the future. *Psychiatric Quarterly, 91*(4), 1121–1133. https://doi.org/10.1007/s11126-020-09808-4. PMID: 32803472; PMCID: PMC7429118.

Ettman, C. K., Abdalla, S. M., Cohen, G. H., Sampson, L., Vivier, P. M., and Galea, S. (2020). Prevalence of depression symptoms in US adults before and during the COVID-19 pandemic. *JAMA Network Open, 3*(9), e2019686. https://doi.org/10.1001/jamanetworkopen.2020.19686.

REFERENCES

Feist, J. B., Feist, J. C., and Cipriano, P. (2020). Stigma Compounds the Consequences of Clinician Burnout During COVID-19: A Call to Action to Break the Culture of Silence. *NAM Perspectives*. Commentary, National Academy of Medicine, Washington, DC. https://doi.org/10.31478/202008b.

Fekedulegn, D., Alterman, T., Charles, L. E., Kershaw, K. N., Safford, M. M., Howard, V. J., and MacDonald, L. A. (2019). Prevalence of workplace discrimination and mistreatment in a national sample of older U.S. workers: The REGARDS cohort study. *SSM Population Health*, 8(July 2, 2019), 100444. https://doi.org/10.1016/j.ssmph.2019.100444. PMID: 31321281; PMCID: PMC6612926.

Felix, E. R., Bensimson, E. M., Hanson, D., Gray, J., and Klingsmith, L. (2015). Developing agency for equity-minded change. *New Directions for Community Colleges*, 172, 25–42.

Fernandez, A. A., and Shaw, G. P. (2020). Academic leadership in a time of crisis: The coronavirus and COVID-19. *Journal of Leadership Studies*, 14(1), 39–45.

Feyerherm, A. E. (1994). Leadership in collaboration: A longitudinal study of two interorganizational rule-making groups. *Leadership Quarterly*, 5(3), 253–270.

Finkelstein, M. J., Conley, V. M., and Schuster, J. H. (2016). *The Faculty Factor: Reassessing the American Academy in a Turbulent Era*. Baltimore, MD: JHU Press.

Flaherty, C. (2020a). "No Room of One's Own: Early Journal Submission Data Suggest COVID 19 is Tanking Women's Research Productivity." *Inside Higher Education*, April 21, 2020. https://www.insidehighered.com/news/2020/04/21/early-journal-submission-data-suggest-COVID-19-tanking-womens-research-productivity.

Flaherty, C. (2020b). "Colleges lower the boom on retirement plans." *Inside Higher Education*, May 21, 2020. https://www.insidehighered.com/news/2020/05/21/more-institutions-are-suspending-or-cutting-retirement-plan-contributions.

Flaherty, C. (2020c). "Mounting faculty concerns about the fall semester." *Inside Higher Education*, June 30, 2020. https://www.insidehighered.com/news/2020/06/30/faculty-concerns-about-fall-are-mounting.

Flaherty, C. (2020d). "Not expendable." *Inside Higher Education*, August 12, 2020. https://www.insidehighered.com/news/2020/08/12/new-aaup-statement-urges-institutions-not-treat-their-adjuncts-expendable.

Flaherty, C. (2020e). "Something's got to give." *Inside Higher Education*, August 20, 2020. https://www.insidehighered.com/news/2020/08/20/womens-journal-submission-rates-continue-fall.

Flaherty, C. (2020f). AAUP to investigate COVID-19 "crisis in academic governance." *Quick Takes*. Inside Higher Education website, September 9, 2020. https://www.insidehighered.com/quicktakes/2020/09/22/aaup-investigate-COVID-19-%E2%80%98crisis-academic-governance%E2%80%99.

Flaherty, C. (2020g). "The souls of Black professors." *Inside Higher Education*, October 21, 2020. https://www.insidehighered.com/news/2020/10/21/scholars-talk-about-being-black-campus-2020.

Fluent, Inc. (2020). Working-from-home perceptions. *Pulse*. https://fluentpulse.com/COVID-19-working-from-home/.

Fond, G., Pauly, V., Leone, M., Llorca, P-M., Orleans, V., Loundou, A., Lancon, C., Auquier, P., Baumstarck, K., and Boyer, L. (2020). Disparities in intensive care unit admission and mortality among patients with schizophrenia and COVID-19: A national cohort study. *Schizophrenia Bulletin*, October 22, 2020, sbaa158. https://doi.org/10.1093/schbul/sbaa158.

Ford, H., Brick, C., Azmitia, M., Blaufuss, K., and Dekens, P. (2019). Women from some underrepresented minorities are given too few talks at the world's largest earth-science conference. *Nature*, 576, 32–35. https://doi.org/10.1038/d41586-019-03688-w.

Foster, D. (2020). *COVID-19 Impact Study: Technical Report. Analysis of the COVID-19 Pivot on the Spring 2020 Administration of the COACHE Faculty Job Satisfaction Survey*. Cambridge, MA: The Collaborative on Academic Careers in Higher Education.

Fox, M. F. (2001). Women, science, and academia: Graduate education and careers. *Gender & Society*, 15(5), 654–666. https://doi.org/10.1177/08912430101500500.

FRB (Federal Reserve Board), Division of Consumer and Community Affairs (DCCA), Consumer and Community Research Section (2020). Update on the Economic Well-Being of U.S. Households: July 2020 Results. September 2020. https://www.federalreserve.gov/publications/files/2019-report-economic-well-being-us-households-update-202009.pdf.

Friga, P. N. (2020). The great recession was bad for higher education. Coronavirus could be worse. *Chronicle of Higher Education*, March 24, 2020. https://www.chronicle.com/article/the-great-recession-was-bad-for-higher-education-coronavirus-could-be-worse/.

Fry, C. V., Cai, X., Zhang, Y., and Wagner, C. S. (2020). Consolidation in a crisis: Patterns of international collaboration in early COVID-19 research. *PLOS ONE* (July 21, 2020). https://doi.org/10.1371/journal.pone.0236307.

Gabster, B. P., van Daalen, K., Dhatt, R., and Barry, M. (2020). Challenges for the female academic during the COVID-19 pandemic. *Lancet*, *395*(10242), 1968–1970.

Galloway, M. K., and Ishimaru, A. M. (2015). Radical recentering: Equity in educational leadership standards. *Educational Administration Quarterly*, *51*(3), 372–408.

Gáti, Á., Tényi, T., Túry, F., and Wildmann, M. (2002). Anorexia nervosa following sexual harassment on the internet: A case report. *International Journal of Eating Disorders*, *31*, 474–477. https://doi.org/10.1002/eat.10029.

Gazerani, P., and Cairns, B. E. (2020). Sex-specific pharmacotherapy for migraine: A narrative review. *Frontiers in Neuroscience*. *14*(March 20), 222. https://doi.org/10.3389/fnins.2020.00222. PMID: 32265634; PMCID: PMC7101090.

Gerull, K. M., Wahba, B. M., Goldin, L. M., McAllister, J., Wright, A., Cochran, A., and Salles, A. (2019). Representation of women in speaking roles at surgical conferences. *American Journal of Surgery*, *220*(1), 20–26. https://doi.org/10.1016/j.amjsurg.2019.09.004.

Gigliotti, R. A. (2019). *Crisis Leadership in Higher Education: Theory and Practice*. New Brunswick, NJ: Rutgers University Press.

Gigliotti, R. (2020). Looking beyond COVID-19: Crisis leadership implications for chairs. *Department Chair*, *31*(1), 14–15.

Gill, R. (2012). "The hidden injuries of the neoliberal university." In *Secrecy and Silence in the Research Process* (pp. 228–244). R. Ryan-Flood & R. Gill (eds.). Routledge.

Ginther, D. K., Currie, J. M., Blau, F. D., and Croson, R. T. A. (2020). Can mentoring help female assistant professors in economics? An evaluation by randomized trial. *American Economic Association Papers and Proceedings*, *110*, 205–209. https://doi.org/10.1257/pandp.20201121.

Glazer-Raymo, J. (2001). *Shattering the Myths: Women in Academe*. Baltimore, MD: Johns Hopkins University Press.

Glynn, S. J. (2018). "An unequal division of labor: How equitable workplace policies would benefit working mothers." Center for American Progress website, May 18, 2020. https://www.americanprogress.org/issues/women/reports/2018/05/18/450972/unequal-division-labor.

Gold, K. J., Andrew, L. B., Goldman, E. B., and Schwenk, T. L. (2016). I would never want to have a mental health diagnosis on my record: A survey of female physicians on mental health diagnosis, treatment, and reporting. *General Hospital Psychiatry*, *43*, 51–57. https://doi.org/10.1016/j.genhosppsych.2016.09.004.

Golden, L. C., and Voskuhl, R. (2017). The importance of studying sex differences in disease: The example of multiple sclerosis. *Journal of Neuroscience Research*, *95*(1-2), 633–643.

Golden, M., Mason, M. A., and Frasch, K. (2011). Keeping women in the science pipeline. *Annals of the American Academy of Political and Social Science*, *638*(1), 141–162. https://doi.org/10.1177/0002716211416925.

Goldin, C. (2014). A grand gender convergence: Its last chapter. *American Economic Review*. *104*(4), 1091–1119. https://doi.org/10.1257/aer.104.4.1091.

Goldin, C., and Mitchell, J. (2017). The new life cycle of women's employment: Disappearing humps, sagging middles, expanding tops. *Journal of Economic Perspectives*, *31*(1), 161–182. http://www.jstor.org/stable/44133995.

REFERENCES

Gonzales, L. D., and Ayers, D. F. (2018). The convergence of institutional logics on the community college sector and the normalization of emotional labor: A new theoretical approach for considering the community college faculty labor expectations. *Review of Higher Education*, *41*(3), 455–478.

Gonzales, L. D., and Griffin, K. A. (2020). *Supporting Faculty During & After COVID-19: Don't Let Go of Equity*. Washington, DC: Aspire Alliance. https://www.mtu.edu/advance/resources/articles-books/supporting-faculty-during-and-after-covid.pdf.

Goodwin, S.A., and Mitchneck, B. (2020). STEM equity and inclusion (Un)interrupted? Inside Higher Ed website, May 13, 2020. https://www.insidehighered.com/views/2020/05/13/ensuring-pandemic-doesnt-negatively-impact-women-stem-especially-those-color.

Gottlieb, M., Egan, D. J., Krzyzaniak, S. M., Wagner, J., Weizberg, M., and Chan, T. (2020) Rethinking the approach to continuing professional development conferences in the era of COVID-19. *Journal of Continuing Education in the Health Professions*, Summer 2020, *40*(3), 187–191. https://doi.org/10.1097/CEH.0000000000000310.

Grandner, M. A., Hale, L., Jackson, N., Patel, N. P., Gooneratne, N. S., and Troxel, W. M. (2012). Perceived racial discrimination as an independent predictor of sleep disturbance and daytime fatigue. *Behavioral Sleep Medicine*, *10*(4), 235–249. https://doi.org/10.1080/15402002.2012.654548. PMID: 22946733; PMCID: PMC3434973.

Gray, D. M., Joseph, J. J., Glover, A. R., and Olayiwola, J. N. (2020). How academia should respond to racism. *Nature Reviews Gastroenterology & Hepatology*, *17*, 589–590. https://doi.org/10.1038/s41575-020-0349-x.

Greenhaus, J. H., and Allen, T. D. (2011). Work-family balance: A review and extension of the literature (pp.165–183), In J. C. Quick and L. E. Tetrick (eds.), *Handbook of Occupational Health Psychology*. Washington, DC: American Psychological Association.

Greenhaus, J. H., and Beutell, N. J. (1985). Sources of conflict between work and family roles. *Academy of Management Review*, *10*(1), 76–88. https://doi.org/10.5465/AMR.1985.4277352.

Greenhaus, J. H., and Powell, G. H. (2006). When work and family are allies: A theory of work-family enrichment. *Academy of Management Review, 31(4)*, 72–92.

Gruber, J., Van Bavel, J. J., Cunningham, W. A., Somerville, L. H., and Lewis Jr., N. A. (2020). Academia needs a reality check: Life is not back to normal. *Science*, August 28, 2020. https://doi.org/10.1126/science.caredit.abe5459.

Guadagni, V., Umilta, A., and Iaria, G. (2020). Sleep quality, empathy, and mood during the isolation period of the COVID-19 pandemic in the Canadian population: Females and women suffered the most. *Frontiers in Global Women's Health*, *1*, 585938. https://doi.org/10.3389/fgwh.2020.585938.

Guan, S., Xiaerfuding, X., Ning, L., Lian, Y., Jiang, Y., Liu, J., and Ng, T. B. (2017). Effect of job strain on job burnout, mental fatigue and chronic diseases among civil servants in the Xinjiang Uygur Autonomous Region of China. *International Journal of Environmental Research and Public Health*, *14*, E872.

Guarino, C. M., and Borden, V. M. (2017). Faculty service loads and gender: Are women taking care of the academic family? *Research in Higher Education*, *58*(6), 672–694. https://doi.org/10.1007/s11162-017-9454-2.

Guatimosim, C. (2020). Reflections on motherhood and the impact of COVID 19 pandemic on women's scientific careers. *Journal of Neurochemistry*. *155*, 469–470. https://doi.org/10.1111/jnc.15158.

Hammer, L. B., Kossek, E. E., Bodner, T., Anger, K., and Zimmerman, K. (2011). Clarifying work-family intervention processes: The roles of work-family conflict and family supportive supervisor behaviors, *Journal of Applied Psychology*, *96*(1), 134–150.

Hanasono, L. K., Broido, E. M., Yacobucci, M. M., Root, K. V., Peña, S., and O'Neil, D. A. (2019). Secret service: Revealing gender biases in the visibility and value of faculty service. *Journal of Diversity in Higher Education*, *12*(1), 85.

Hansen, D. S. (2020). Identifying barriers to career progression for women in science: Is COVID-19 creating new challenges? *Trends in Parasitology, 36*(10), 799–802.

Harned, M. S., and Fitzgerald, L. F. (2002). Understanding a link between sexual harassment and eating disorder symptoms: A mediational analysis. *Journal of Consulting and Clinical Psychology, 70*, 1170–1181.

Harris, J. C., and Patton, L. D. (2019). Un/Doing intersectionality through higher education research. *Journal of Higher Education, 90*(3), 347–372. https://doi.org/10.1080/00221546.2018.1536936.

Harry, E. M., Sinsky, C., Dyrbye, L., Hamidi, M., Trockel, M., Tutty, M., Carlasare, L., West, C. P., and Shanafelt, T. (2019). Physician task load and the risk of burnout among US physicians in a national survey. Paper presented at the Society for Hospital Medicine Annual Meeting, Washington, D.C.

Hart, S. G., and Staveland, L. E. (1988). Development of NASA-TLX (Task Load Index): Results of empirical and theoretical research. In P. A. Hancock and N. Meshkati (eds.), *Advances in Psychology, Volume 52, Human Mental Workload*, 139–183. Amsterdam, NL: Elsevier. https://doi.org/10.1016/S0166-4115(08)62386-9.

Hecht, T. D., and Allen, N. J. (2009). A longitudinal examination of the work–nonwork boundary strength construct. *Journal of Organizational Behavior, 30*(7), 839–862. https://doi.org/10.1002/job.579.

Heggeness, M. L. (2020). Estimating the immediate impact of the COVID-19 shock on parental attachment to the labor market and the double bind of mothers. *Review of Economics of the Household, 18*, 1053–1078. https://doi.org/10.1007/s11150-020-09514-x.

Heidt, A. (2020). Coronavirus precautions stall Antarctic field research. *Scientist*, June 15, 2020. https://www.the-scientist.com/news-opinion/coronavirus-precautions-stall-antarctic-field-research-67636.

Heifetz, R. A. (1994). *Leadership Without Easy Answers*. Cambridge, MA: Harvard University Press.

Hemelt, S. W., and Stange, K. M. (2020). "Why the Move to Online Instruction Won't Reduce College Costs." *Brown Center Chalkboard,* July 28, 2020. Washington, DC: Brookings Institution. https://www.brookings.edu/blog/brown-center-chalkboard/2020/07/28/why-the-move-to-online-instruction-wont-reduce-college-costs/.

Herrero San Martin, A., Parra Serrano, J., Diaz Cambriles, T., Arias Arias, E.M., Muñoz Méndez, J., del Yerro Álvarez , M.J., and González Sánchez, M.. (2020). Sleep characteristics in health workers exposed to the COVID-19 pandemic. *Sleep Medicine*. 75(November), 388-394. doi: 10.1016/j.sleep.2020.08.013.

Hochschild, A. R. (1989). *The Second Shift. Working Parents and the Revolution at Home* (2nd Ed.). New York: Viking Press.

Holman, L., Stuart-Fox, D., and Hauser, C. E. (2018). The gender gap in science: How long until women are equally represented? *PLOS Biology, 16*(4), e2004956.

Holshue, M., Debolt, C., Lindquist, S., Lofy, K., Wiesman, J., Bruce, H., Spitters, C., Ericson, K., Wilkerson, S., Tural, A., Diaz, G., Cohn, A., Fox, L., Patel, A., Gerber, S., Kim, L., Tong, S., Lu, X., Lindstrom, S., Pallansch, M., Weldon, W., Biggs, H., Uyeki, T., and Pillai, S. (2020). First case of 2019 novel coronavirus in the United States. *New England Journal of Medicine, 382*, 929–936. https://doi.org/10.1056/NEJMoa2001191. https://www.nejm.org/doi/full/10.1056/NEJMoa2001191.

Howe-Walsh, L., and Turnbull, S. (2016). Barriers to women leaders in academia: Tales from science and technology. *Studies in Higher Education, 41*(3), 415–428.

Hoynes, H., Miller, D. L., and Schaller, J. (2012). Who suffers during recessions? *Journal of Economic Perspectives, 26*(3), 27–48.

Hrabowski III, F. A. (2019). *The Empowered University: Shared Leadership, Culture Change, and Academic Success*. Baltimore, MD: Johns Hopkins University Press.

Iaria, A., Schwarz, C., and Waldinger, F. (2018). Frontier knowledge and scientific production: Evidence from the collapse of international science. *Quarterly Journal of Economics*, 927–991. https://doi.org/10.1093/qje/qjx046.

REFERENCES

Ibarra, H., Carter, N., and Silva, C. (2010). Why men still get more promotions than women. *Harvard Business Review, 88*(September), 1–6.

IBM Corp. (2019). *IBM SPSS Statistics for Windows, Version 26.0*. Armonk, NY: IBM Corp.

ILO (International Labor Organization) (2020). Policy brief. The COVID-19 response: Getting gender equality right for a better future for women at work. Geneva: International Labor Organization. https://www.ilo.org/global/topics/coronavirus/WCMS_744685/lang--en/index.htm.

Inouye, D. W., Underwood, N., Inouye, B. D., and Irwin, R. E. (2020). Support early-career field researchers. *Science, 368*(6492), 724–725.

IOM (Institute of Medicine) (2011). *The Health of Lesbian, Gay, Bisexual, and Transgender People: Building a Foundation for Better Understanding*. Washington, DC: The National Academies Press. PMID: 22013611.

Jagsi, R., Griffith, K. A., Jones, R., Perumalswami, C. R., Ubel, P., and Stewart, A. (2016). Sexual harassment and discrimination experiences of academic medical faculty. *JAMA, 315*(19), 2120–2121. https://doi.org/10.1001/jama.2016.2188.

Jagsi, R., Jones, R. D., Griffith, K. A., Brady, K. T., Brown, A. J., Davis, R. D., Drake, A. F., Ford, D., Fraser, V. J., Hartmann, K. E., Hochman, J. S., Girdler, S., Libby, A. M., Mangurian, C., Regensteiner, J. G., Yonkers, K., Escobar-Alvarex, S., and Myers, E. R. (2018). An innovative program to support gender equity and success in academic medicine: Early experiences from the Doris Duke Charitable Foundation's Fund to Retain Clinical Scientists. *Annals of Internal Medicine, 169*(2), 128–130. https://doi.org/10.7326/M17-2676. Epub 2018 Mar 20.PMID: 29554690.

Jensen, L. E., and Deemer, E. D. (2019). Identity, campus climate, and burnout among undergraduate women in STEM fields. *Career Development Quarterly, 67*(2), 96–109. https://doi.org/10.1002/cdq.12174.

Johnson, K. A. C. (2018). 9/11 and international student visa issuance. *Journal of Studies in International Education, 22*(5), 393–413. https://doi.org/10.1177/1028315318762524.

Jolly, S., Griffith, K. A., DeCastro, R. Stewart, A., Ubel, P., and Jagsi, R. (2014). Gender differences in time spent on parenting and domestic responsibilities by high-achieving young physician-researchers. *Annals of Internal Medicine, 160*(5), 344–353. https://doi.org/10.7326/M13-0974.

Jones, R. D., Miller, J., Vitous, C. A., Krenz, C., Brady, K. T., Brown A. J., Daumit, G. L., Drake, A.F., Fraser, V. J., Hartmann, K. E., Hochman, J. S., Girdler, S., Kalet, A.L., Libby, A.M., Mangurian, C., Regensteiner, J.G., Yonkers, K., and Jagsi, R. (2020). From stigma to validation: A qualitative assessment of a novel national program to improve retention of physician-scientists with caregiving responsibilities. *Journal of Women's Health (Larchmt), 29*(12), 1547–1558. https://doi.org/10.1089/jwh.2019.7999.

Jones, R. D., Miller, J., Vitous, C. A., Krenz, C., Brady, K. T., Brown, A. J., Daumit, G. L., Drake, A. F., Fraser, V. J., Hartmann, K. E., Hochman, J. S., Girdler, S., Libby, A. M., Mangurian, C., Regensteiner, J. G., Yonkers, K., and Jagsi, R. (2019). The most valuable resource is time: Insights from a novel national program to improve retention of physician-scientists with caregiving responsibilities. *Academic Medicine, 94*(11), 1746–1756. https://doi.org/10.1097/ACM.0000000000002903. PMID: 31348060.

Jorgenson, J. (2002). Engineering selves negotiating gender and identity in technical work. *Management Communication Quarterly, 15*(3), 350–380. https://doi.org/10.1177/0893318902153002.

Jostell, D., and Hemlin, S. (2018). After hours teleworking and boundary management: Effects on work-family conflict. *Work, 60*(3), 475–483. https://doi.org/10.3233/WOR-182748.

Kahn, L. B. (2010). The long-term labor market consequences of graduating from college in a bad economy. *Labour Economics, 17*(2), 303–316. https://doi.org/10.1016/j.labeco.2009.09.002.

Kahn, R. L., Wolfe, D. M., Quinn, R. P., Snoek, J. D., and Rosenthal, R. A. (1964). *Organizational Stress: Studies in Role Conflict and Ambiguity*. New York: Wiley.

Kalev, A. (2020). "U.S. Unemployment Rising Faster for Women and People of Color." *Harvard Business Review*, April 20, 2020. https://hbr.org/2020/04/research-u-s-unemployment-rising-faster-for-women-and-people-of-color.

Kalia, V., Srinivasan, A., Wilkins, L., and Luker, G. D. (2020). Adapting scientific conferences to the realities imposed by COVID-19. *Radiology Imaging Cancer*, 2(4), e204020. https://doi.org/10.1148/rycan.2020204020.

Kannampallil, T. G., Goss, C. W., Evanoff, B. A., Strickland, J. R., McAlister, R. P., and Duncan, J. (2020). Exposure to COVID-19 patients increases physician trainee stress and burnout. *PLOS ONE*, 15(8), e0237301. https://doi.org/10.1371/journal.pone.0237301.

Kanter, R. M. (1977). *Work and Family in the United States: A Critical Review and Future Agenda for Research and Policy.* New York: Russell Sage Foundation.

Katz, D., and Kahn, R. (1966). *The Social Psychology of Organizations.* New York: Wiley.

Kawaguchi, D., and Kondo, A. (2019). The effects of graduating from college during a recession on living standard, *Economic Inquiry*, 58(1), 283–293. https://doi.org/10.1111/ecin.12835.

Kendi, I. X. (2019). *How to Be an Antiracist.* New York: Random House.

Kent, D. G., Knapp, D. J. H. F., and Kannan, N. (2020). Survey says: "COVID-19 lockdown hits young faculty and clinical trials." *Stem Cell Reports*, 15(1), 1–5. https://doi.org/10.1016/j.stemcr.2020.06.010.

Kezar, A., DePaola, T., and Scott, D. T. (2019). *The Gig Academy: Mapping Labor in the Neoliberal University.* Baltimore, MD: Johns Hopkins University Press.

Kezar, A., Fries-Britt, S., Kurban, E., McGuire, D., and Wheaton, M. M. (2018). *Speaking Truth and Acting with Integrity: Confronting Challenges of Campus Racial Climate.* Washington, DC: American Council on Education.

Kezar, A., and Holcombe, E. (2017). *Shared Leadership in Higher Education: Important Lessons from Research and Practice.* Washington, DC: American Council on Education.

Kezar, A., and Posselt, J. (2020). *Higher Education Administration for Social Justice and Equity: Critical Perspectives for Leadership.* New York: Routledge.

Khatony, A., Zakiei, A., Khazaie, H., Rezaei, M., and Janatolmakan, M. (2020). International nursing: A study of sleep quality among nurses and its correlation with cognitive factors. *Nursing Administration Quarterly*, 44(1), E1–E10. https://doi.org/10.1097/NAQ.0000000000000397. PMID: 31789753.

Kibbe, M. R. (2020). Consequences of the COVID-19 pandemic on manuscript submissions by women. *JAMA Surgery*, 155(9), 803–804.

Kim, E. J., and Dimsdale, J. E. (2007). The effect of psychosocial stress on sleep: A review of polysomnographic evidence. *Behavioral Sleep Medicine*, 5(4), 256–278. https://doi.org/10.1080/15402000701557383.

Kobayashi, I., and Mellman, T. A. (2012). Gender differences in sleep during the aftermath of trauma and the development of posttraumatic stress disorder. *Behavioral Sleep Medicine*, 10(3), 180–90. https://doi.org/10.1080/15402002.2011.654296. PMID: 22742436; PMCID: PMC3947587.

Kochhar, R. (2020). Fewer mothers and fathers in U.S. are working due to COVID-19 downturn; those at work have cut hours. *Fact Tank*, posted October 22, 2020. Pew Research Center. https://www.pewresearch.org/fact-tank/2020/10/22/fewer-mothers-and-fathers-in-u-s-are-working-due-to-COVID-19-downturn-those-at-work-have-cut-hours/.

Kondo, A. (2015). Differential effects of graduating during a recession across gender and race. *IZA Journal of Labor Economics*, 4(23). https://doi.org/10.1186/s40172-015-0040-6.

Korbel, J. O., and Stegle, O. (2020). Effects of the COVID-19 pandemic on life scientists. *Genome Biology*, 21, 113. https://doi.org/10.1186/s13059-020-02031-1.

Kossek, E. E. (2005). Workplace policies and practices to support work and families. In S. Bianchi, L. Casper, and R. King (eds.), *Work, Family, Health, and Well-Being*, 97–116. Mahwah, NJ: Lawrence Erlbaum Associates, Press.

Kossek, E. E. (2006a). Work and family in America: Growing tensions between employment policy and a changing workforce. A thirty-year perspective. In E. Lawler and J. O'Toole (eds.), *America at Work: Choices and Challenges*, 53–72. New York: Palgrave MacMillan.

REFERENCES

Kossek, E. E. (2006b). Changing workforce. A thirty-year perspective. (Commissioned chapter by SHRM Foundation and University of California Center for Organizational Effectiveness for the 30th anniversary of the State of Work in America.) In E. Lawler and J. O'Toole (eds.), *America at Work: Choices and Challenges*, 53–72. New York: Palgrave MacMillan.

Kossek, E. E. (2016). Managing work–life boundaries in the digital age. *Organizational Dynamics*, 45(3), 258–270. http://dx.doi.org/10.1016/j.orgdyn.2016.07.010.

Kossek, E. E. (2020). Exploring an organizational science view on faculty gender and work-life inclusion: Conceptualization, perspectives, and interventions. In E. Kossek and K.-H. Lee (eds.), *Fostering Gender and Work-Life Inclusion for Faculty in Understudied Contexts: An Organizational Science Lens*, 28–32. West Lafayette, IN: Purdue e-Pubs. https://doi.org/10.5703/1288284317255. https://docs.lib.purdue.edu/worklifeinclusion/2018/gwlibsufc/3/.

Kossek, E. E., Allen, T., and Lee, K.-H. (2019–2021). Faculty Work-Life Boundary Management, Inclusion, and Women's Career Well-being in the Always-On Workplace: A National Survey (NSF # 1922380; PI: Kossek, E. E.). https://www.nsf.gov/awardsearch/showAward?AWD_ID=1922380&HistoricalAwards=false.

Kossek, E. E., Colquitt, J., and Noe, R. A. (2001). Caregiving decisions, well-being and performance: The effects of place and provider as a function of dependent type and work-family climates. *Academy of Management Journal*, 44(1), 29–44. https://doi.org/10.2307/3069335.

Kossek, E., Hammer, L., Kelly, E., and Moen, P. (2014). Designing organizational work, family and health change initiatives. *Organizational Dynamics*, 43, 53–63.

Kossek, E., and Lambert, S. (eds.) (2005). "Introduction" *Work and Life Integration: Organizational, Cultural, and Individual Perspectives*. Mahwah, NJ: Lawrence Erlbaum Associates.

Kossek, E., Lautsch, B., and Eaton, S. (2006). Telecommuting, control, and boundary management: Correlates of policy use and practice, job control, and work-family effectiveness. *Journal of Vocational Behavior*, 68(2), 347–367. https://doi.org/10.1016/j.jvb.2005.07.002.

Kossek, E. E., and Lautsch, B. A. (2012). Work family boundary management styles in organizations A cross-level model. *Organizational Psychology Review*, 2(2), 152–171. https://doi.org/10.1177/2041386611436264.

Kossek, E., and Lautsch, B. (2018). Work-life flexibility for whom? Occupational status and work-life inequality in upper, middle, and lower level jobs. *Academy of Management Annals*, 12(1), 5–36.

Kossek, E. E., and Lee, K.-H. (2020a). *Fostering Gender and Work-Life Inclusion for Faculty in Understudied Contexts: An Organizational Science Lens*. West Lafayette, IN: Purdue ePubs. https://docs.lib.purdue.edu/worklifeinclusion/worklifeinclusion_monograph.pdf.

Kossek, E. E., and Lee, K.-H. (2020b). The coronavirus & work–life inequality: Three evidence-based initiatives to update U.S. work–life employment policies. *Behavioral Science & Policy*, COVID-19 Special Issue. https://behavioralpolicy.org/journal_issue/COVID-19/.

Kossek, E. E., and Lee, K. H. (2020c). Work-life Inclusion for Women's Career Equality: Why it Matters and What to Do About It. Working paper. West Lafayette, IN: Purdue University.

Kossek, E. E., Lewis, S., and Hammer, L. (2010). Work-life initiatives and organizational change: Overcoming mixed messages to move from the margin to the mainstream. *Human Relations*, 63, 1–17. https://doi.org/10.1177/0018726709352385.

Kossek, E. E., and Ollier-Malaterre, A. (2019). Desperately seeking sustainable careers: Redesigning professional jobs for the collaborative crafting of reduced-load work. *Journal of Vocational Behavior*, 117. https://doi.org/10.1016/j.jvb.2019.06.003.

Kossek, E. E., and Ozeki, C. (1998). Work- family conflict, policies, and the job-life satisfaction relationship: A review and directions for organizational behavior/human resources research. *Journal of Applied Psychology*, 83(2), 139–149.

Kossek, E., Perrigino, M., and Gounden- Rock, A. (forthcoming). From ideal workers to ideal work for all: A 50-year review integrating the careers and work-family literatures. *Journal of Vocational Behavior*.

Kossek, E., Pichler, S., Bodner, T., and Hammer, L. (2011). Workplace social support and work-family conflict: A meta-analysis clarifying the influence of general and work-family specific supervisor and organizational support. *Personnel Psychology, 64*(2), 289–313.

Kossek, E. E., Ruderman, M., Braddy, P., and Hannum, K. (2012). Work-nonwork boundary management profiles: A person-centered approach, *Journal of Vocational Behavior, 81*, 112–128. http://dx.doi.org/10.1016/j.jvb.2012.04.003.

Kossek, E. E., Thompson, R. J., Lawson, K. M., Bodner, T. Perrigino, M., Hammer, L. B., Buxton, O. M., Almeida, D. M., Moen, P., Hurtado, D., Wipfli, B., Berkman, L. F., and Bray, J. W. (2019). Caring for the elderly at work and home: Can a randomized organizational intervention improve psychological health? *Journal of Occupational Health Psychology, 24*(1), 36–54. http://dx.doi.org/10.1037/ocp0000104 https://www.ncbi.nlm.nih.gov/pubmed/29215909.

Kram, K. (1985). *Mentoring at Work: Developmental Relationships in Organizational Life*. Glenview, IL: Scott Foresman.

Kramer, J. (2020a). Women in science may suffer lasting career damage from COVID-19. *Scientific American*, August 12, 2020. https://www.scientificamerican.com/article/women-in-science-may-suffer-lasting-career-damage-from-COVID-19/.

Kramer, J. (2020b). "The Virus Moved Female Faculty to the Brink. Will Universities Help?" *New York Times*, October 6, 2020. https://www.nytimes.com/2020/10/06/science/covid-universities-women.html.

Kreeger, P. K., Brock, A., Gibbs, H. C., Grande-Allen, K. J., Huang, A. H., Masters, K. S., and Servoss, S. L. (2020). Ten simple rules for women principal investigators during a pandemic. *PLOS Computational Biology, 16*(10), e1008370.

Kreiner, G. E. (2006). Consequences of work-home segmentation or integration: A person-environment fit perspective. *Journal of Organizational Behavior, 27*(4), 485–507. https://doi.org/10.1002/job.386.

Krishnamoorthy, Y., Nagarajan, R., Saya, G. K., and Menon, V. (2020). Prevalence of psychological morbidities among general population, healthcare workers and COVID-19 patients amidst the COVID-19 pandemic: A systematic review and meta-analysis. *Psychiatry Research*, 293, 113382. https://doi.org/10.1016/j.psychres.2020.113382.

Kristensen, T., Borritz, M., Villadsen, E., and Christensen, K. (2005). The Copenhagen Burnout Inventory: A new tool for the assessment of burnout. *Work & Stress, 19*(3), 192–207. https://doi.org/10.1080/02678370500297720.

Kroenke, K., Spitzer, R. L., and Williams, J. B. (2001). The PHQ-9: validity of a brief depression severity measure. *Journal of General Internal Medicine, 16*(9), 606–613. https://doi.org/10.1046/j.1525-1497.2001.016009606.x. PMID: 11556941; PMCID: PMC1495268.

Kroenke, K., Spitzer, R. L., and Williams, J. B. (2003). The Patient Health Questionnaire-2: Validity of a two-item depression screener. *Medical Care, 41*(11), 1284–1292. https://doi.org/10.1097/01.MLR.0000093487.78664.3C. PMID: 14583691.

Krueger, C., and Tian, L. (2004). A comparison of the general linear mixed model and repeated measures ANOVA using a dataset with multiple missing data points. *Biological Research for Nursing, 6*(2), 151–157. https://doi.org/10.1177/1099800404267682.

Krukowski, R. A., Jagsi, R., and Cardel, M. I. (2020). Academic productivity differences by gender and child age in science, technology, engineering, mathematics, and medicine faculty during the COVID-19 pandemic. *Journal of Women's Health*, published online November 18, 2020. https://doi.org/10.1089/jwh.2020.8710.

Lai, J., Ma, S., Wang, Y., Cai, Z., Hu, J., Wei, N., Wu, J., Du, H., Chen, T., Li, R., Tan, H., Kang, L., Yao, L., Huang, M., Wang, H., Wang, G., Liu, Z., and Hu, S. (2020). Factors associated with mental health outcomes among health care workers exposed to coronavirus disease 2019. *JAMA Network Open, 3*(3), e203976. https://doi.org/10.1001/jamanetworkopen.2020.3976. PMID: 32202646; PMCID: PMC7090843.

Laner, M. R. (2000). "Sex" versus "gender": A renewed plea. *Sociological Inquiry, 70*(4), 462–474. doi:10.1111/j.1475-682X.2000.tb00920.x.

REFERENCES

Langan, A. (2019). Female managers and gender disparities: The case of academic department chairs. *Job Market Paper*, Princeton University website. https://scholar.princeton.edu/sites/default/files/alangan/files/langan_jmp_current.pdf.

Lapierre, L. M., Steenbergen, E. F., Peeters, M. C. W., and Kluwer, E. S. (2016). Juggling work and family responsibilities when involuntarily working more from home: A multiwave study of financial sales professionals. *Journal of Organizational Behavior*, *37*(6), 804–822. https://doi.org/10.1002/job.2075.

Larson, A. R., Sharkey, K. M., Poorman, J. A., Kan, C. K., Moeschler, S. M., Chandrabose, R., Marquez, C. M., Dodge, D. G., Silver, J. K., and Nazarian, R. M. (2019). Representation of women among invited speakers at medical specialty conferences. *Journal of Women's Health*, *29*(4), 550–560. https://doi.org/10.1089/jwh.2019.7723.

Levine, C., Miller, E. C., Morgan, B., Taylor, J. O., Voskuil, L., Williams, D., Kreisel, D., Roop, H., Wheeler, S., Bloom, A., Howart, R. W., and Jonsson, F. A. (2019). Reducing the carbon footprint of academic travel. Inside Higher Education website, April 18, 2019. https://www.insidehighered.com/views/2019/04/18/12-scholars-share-ideas-reducing-carbon-emissions-academic-travel-opinion.

Li, L., Wu, C., Gan, Y., Qu, X., and Lu, Z. (2016). Insomnia and the risk of depression: A meta-analysis of prospective cohort studies. *BMC Psychiatry*, *16*(1), 375. https://doi.org/10.1186/s12888-016-1075-3. PMID: 27816065; PMCID: PMC5097837.

Linos, E., Halley, M., Sarkar, U., Mangurian, C., Sabry, H., Olazo, K., Mathews, K. S., Diamond, L., Goyal, M. K., Linos, E., and Jagsi, R. (2020). "Letter to the Editor: Anxiety Levels among Physician-Mothers during the COVID Pandemic. https://ajp.psychiatryonline.org/pb-assets/journals/ajp/homepage/Anxiety%20Levels%20Among%20Physician-Mothers%20During%20the%20COVID%20Pandemic.pdf.

Linzer, M., Konrad, T. R., Douglas, J., McMurray, J. E., Pathman, D. E., Williams, E. S., Schwartz, M. D., Gerrity, M., Scheckler, W., Bigby, J. A., and Rhodes, E. (2000). Managed care, time pressure, and physician job satisfaction: Results from the physician worklife study. *Journal of General Internal Medicine*, *15*(7), 441–450. https://doi.org/10.1046/j.1525-1497.2000.05239.x. PMID: 10940129; PMCID: PMC1495485.

Liu, N., Zhang, F., Wei, C., Jia, Y., Shang, Z., Sun, L., Wu, L., Sun, Z., Zhou, Y., Wang, Y., and Liu, W. (2020a). Prevalence and predictors of PTSS during COVID-19 outbreak in China hardest-hit areas: Gender differences matter. *Psychiatry Research*, *287*, 112921. https://doi.org/10.1016/j.psychres.2020.112921.

Liu, R. T., Steele, S. J., Hamilton, J. L., Do, Q. B. P., Furbish, K., Burke, T. A., Martinez, A. P., and Gerlus, N. (2020b). Sleep and suicide: A systematic review and meta-analysis of longitudinal studies. *Clinical Psychology Review*, *81*, 101895. https://doi.org/10.1016/j.cpr.2020.101895. Epub August 8, 2020. PMID: 32801085.

Lu, D. W., Lall, M. D., Mitzman, J., Heron, S., Pierce, A., Hartman, N. D., McCarthy, D. M., Jauregui, J., and Strout, T. D. (2020). #MeToo in EM: A multicenter survey of academic emergency medicine faculty on their experiences with gender discrimination and sexual harassment. *Western Journal of Emergency Medicine*, *21*(2), 252–260. https://doi.org/10.5811/westjem.2019.11.44592. PMID: 32191183; PMCID: PMC7081862.

Lyu, H. G., Davids, J. S., Scully, R. E., and Melnitchouk, N. (2019). Association of domestic responsibilities with career satisfaction for physician mothers in procedural vs nonprocedural fields. *JAMA Surgery*, *154*(8), 689–695. https://doi.org/10.1001/jamasurg.2019.0529.

Madgavkar, A., White, O., Krishnan, M., Mahajan, D., and Azcue, X. (2020). COVID-19 and gender equality: Countering the regressive effects. McKinsey & Company website, Article, July 15, 2020. https://www.mckinsey.com/featured-insights/future-of-work/COVID-19-and-gender-equality-countering-the-regressive-effects#.

Madsen, T. E., Dobiesz, V., Das, D., Sethuraman, K., Agrawal, P., Zeidan, A., Goldberg, E., Safdar, B., and Lall, M. D. (2020). Unique risks and solutions for equitable advancement during the COVID-19 pandemic: Early experience from frontline physicians in academic medicine. *NEJM Catalyst Innovations in Care Delivery*, *1*(4).

Malisch, J. L., Harris, B. N., Sherrer, S. M., Lewis, K. A., Shepherd, S. L., Pumtiwitt, C., McCarthy, P. C., Spott, J. L., Karam, E. P., Moustaid-Moussa, N., McCrory Calarco, J., Ramalingam, L., Talley, A. E., Jaclyn, E., Cañas-Carrell, J. E., Ardon-Dryer, K., Weiser, D. A., Bernal, X. E., and Deitloff, J. (2020). Opinion: In the wake of COVID-19, academia needs new solutions to ensure gender equity. *Proceedings of the National Academy of Sciences of the United States of America, 117*(27), 15378–15381. https://doi.org/10.1073/pnas.2010636117.

Mangrio, E., and Sjögren Forss, K. (2017). Refugees' experiences of healthcare in the host country: a scoping review. *BMC Health Services Research, 17*(1), 814. https://doi.org/10.1186/s12913-017-2731-0. PMID: 29216876; PMCID: PMC5721651.

Mangurian, C., Linos, E., Sarkar, U., Rodriguez, C., and Jagsi, R. (2018). What's holding women in medicine back from leadership. *Harvard Business Review*, June 19, 2018, updated November 7, 2018. https://hbr.org/2018/06/whats-holding-women-in-medicine-back-from-leadership.

Maranto, C. L., and Griffin, A. E. (2011). The antecedents of a "chilly climate" for women faculty in higher education. *Human Relations, 64*(2), 139–159. https://doi.org/10.1177/0018726710377932.

Marelli, S., Castelnuovo, A., Somma, A., Castronovo, V., Mombelli, S., Bottoni, D., Leitner, C., Fossati, A., and Ferini-Strambi, L. (2020). Impact of COVID-19 lockdown on sleep quality in university students and administration staff. *Journal of Neurology*, July 11, 2020, 1–8. https://doi.org/10.1007/s00415-020-10056-6. Epub ahead of print. PMID: 32654065; PMCID: PMC7353829.

Maslach, C., and Jackson, S.E. (1993). Maslach Burnout Inventory Manual. 2nd ed. Palo Alto, CA: Consulting Psychologists Press.

Matthews, R. A. (2020). Barriers to Organizational Work-Family Support in Academia: An HR perspective. In E. Kossek and K.-H. Lee (eds.), *Fostering Gender and Work-Life Inclusion for Faculty in Understudied Contexts: An Organizational Science Lens*, 110–115. West Lafayette, IN: Purdue e-Pubs. https://doi.org/10.5703/1288284317222. https://docs.lib.purdue.edu/worklifeinclusion/2018/dccsowfs/2.

Mavriplis, C., Heller, R., Beil, C., Dam, K., Yassinskaya, N., Shaw, M., and Sorensen, C. (2010). Mind the gap: Women in STEM career breaks. *Journal of Technology Management and Innovation, 5*(1), 140–151.

McAuliff, M., Valdivia, S. M., Herman, C., and O'Neill, S. (2020). Swab, spit or stay home? A wide variety of plans to keep coronavirus off campus. *National Public Radio*, August 20, 2020. https://www.npr.org/sections/health-shots/2020/08/20/904114299/swab-spit-or-stay-home-a-wide-variety-of-plans-to-keep-coronavirus-off-campus.

McCoy, H. (2020). "The Life of a Black Academic: Tired and Terrorized", Inside Higher Education website, June 12, 2020. https://www.insidehighered.com/advice/2020/06/12/terror-many-black-academics-are-experiencing-has-left-them-absolutely-exhausted.

McKay, D., and Asmundson, G. J. (2020). COVID-19 stress and substance use: Current issues and future preparations. *Journal of Anxiety Disorders*, July 21, 2020.

McMurtrie, B. (2020). "The pandemic is dragging on. Professors are burning out. Overwhelmed and unsupported, instructors see no end in sight." *Chronicle of Higher Education*, November 5, 2020. https://www.chronicle.com/article/the-pandemic-is-dragging-on-professors-are-burning-out.

McSwain, C. (2019). "Seeking Diversity vs. Anti-Racism – What's the difference?" National Juvenile Justice Network, April 26, 2019. https://njjn.org/article/seeking-diversity-vs--anti-racism---what's-the-difference.

Mendoza-Denton, R., Patt, C., Fisher, A., Eppig, A., Young, I., Smith, A., and Richards, M. A. (2017). Differences in STEM doctoral publication by ethnicity, gender and academic field at a large public research university. *PLOS ONE, 12*(4), e0174296.

Mensah, M., Beeler, W., Rotenstein, L., Jagsi, R., Spetz, J., Linos, E., and Mangurian, C. (2020). Sex differences in salaries of department chairs at public medical schools. *JAMA Internal Medicine, 180*(5), 789–792. https://doi.org/10.1001/jamainternmed.2019.7540.

REFERENCES

Messias, E., Gathright, M. M., Freeman, E. S., Flynn, V., Atkinson, T., Thrush, C. R., Clardy, J. A., and Thapa, P. (2019). Differences in burnout prevalence between clinical professionals and biomedical scientists in an academic medical centre: A cross-sectional survey. *BMJ Open*, 9(2), e023506. https://doi.org/10.1136/bmjopen-2018-023506. PMID: 30782882; PMCID: PMC6367953.

Metcalfe, A., and Slaughter, S. (2008). The differential effects of academic capitalism on women in the academy. In J. Glazer-Raymo (ed.), *Unfinished Agendas: New and Continuing Gender Challenges in Higher Education* (pp. 80–111). Baltimore, MD: Johns Hopkins University.

Mickey, E. L. (2019). STEM Faculty Networks and Gender: A Meta-Analysis. ARC Network/Association for Women in Science, July 15, 2019. https://www.equityinstem.org/wp-content/uploads/EMickey-STEM-Faculty-Networks-Gender-White-Paper.pdf.

Mickey, E. L., Kanelee, E. S., and Misra, J. (2020). 10 small steps for department chairs to foster inclusion. Inside Higher Education website, June 5, 2020. https://www.insidehighered.com/advice/2020/06/05/advice-department-chairs-how-foster-inclusion-among-faculty-opinion.

Miller, C. C. (2020). "Nearly Half of Men Say They Do Most of the Home Schooling. 3 Percent of Women Agree." *The New York Times*, May 6, 2020. https://www.nytimes.com/2020/05/06/upshot/pandemic-chores-homeschooling-gender.html.

Minello, A. (2020). The pandemic and the female academic. *Nature*, April 17, 2020. https://www.nature.com/articles/d41586-020-01135-9.

Minor, J. T. (2008). *Contemporary HBCUs: Considering Institutional Capacity and State Priorities. A Research Report.* East Lansing, MI: Department of Educational Administration, College of Education, Michigan State University.

Misra, J., Lundquist, J., and Templar, A. (2012). Gender, work time and care responsibilities among faculty. *Sociological Forum*, 27(2), 300–323. https://doi.org/10.1111/j.1573-7861.2012.01319.x.

Moore, H.A., Acosta, K., Perry, G., and Edwards, C. (2010). Splitting the academy: The emotions of intersectionality at work. *Sociological Quarterly*, 51(2), 179–204.

Mor Barak, M. E. (2020). Blind spots in work-life research through a global lens: Toward a model of intersectionality, diversity and inclusion. In E. Kossek and K.-H. Lee (eds.), *Fostering Gender and Work-Life Inclusion for Faculty in Understudied Contexts: An Organizational Science Lens*, 34–40. West Lafayette, IN: Purdue e-Pubs. https://doi.org/10.5703/1288284317258. https://docs.lib.purdue.edu/worklifeinclusion/2018/idgwli/1/.

Morin, C. M., Belleville, G., Bélanger, L., and Ivers, H. (2011). The insomnia severity index: Psychometric indicators to detect insomnia cases and evaluate treatment response. *Sleep*, 34(5), 601–608. https://doi.org/10.1093/sleep/34.5.601.

Morris, V. R., and Washington, T. M. (2018). The role of professional societies in STEM diversity. *Journal of the National Technical Association*, 87(1), 22–31. http://dx.doi.org/10.1090/noti1642.

Moss, P., Salzman, H., and Tilly, C. (2005). When firms restructure: Understanding work-life outcomes. In E. Kossek and S. Lambert (eds.), *Work and Life Integration: Organizational, Cultural, and Individual Perspectives*. Mahwah, NJ: Lawrence Erlbaum Associates.

Mullangi, S., Blutt, M. J., and Ibrahim, S. (2020). Is it time to reimagine academic promotion and tenure? *JAMA Health Forum*, published online February 25, 2020. DOI:10.1001/jamahealthforum.2020.0164.

Muric, G., Lerman, K., and Ferrara, E. (2020). COVID-19 amplifies gender disparities in research. *arXiv*, 2006.06142. https://arxiv.org/abs/2006.06142.

Myers, K. R., Tham, W. Y., Yin, Y., Cohodes, J., Thursby, J. G., Thursby, M. C., Schiffer, P., Walsh, J. T., Lakhani, K. R., and Wang, D. (2020). Unequal effects of the COVID-19 pandemic on scientists. *Nature Human Behavior*, 4, 880–883. https://doi.org/10.1038/s41562-020-0921-y.

NAE (National Academy of Engineering) (2017). *Engineering Societies and Undergraduate Engineering Education: Proceedings of a Workshop*. Washington, DC: The National Academies Press. https://doi.org/10.17226/24878.

NAM (National Academy of Medicine) (n.d.) *Valid and Reliable Survey Instruments to Measure Burnout, Well-Being, and Other Work-Related Dimensions.* Washington, DC: The National Academies Press. https://nam.edu/valid-reliable-survey-instruments-measure-burnout-well-work-related-dimensions/.

NASA (National Aeronautics and Space Administration) (n.d.). Technical Reports Server. Task load index. https://ntrs.nasa.gov/archive/nasa/casi.ntrs.nasa.gov/20000021488.pdf.

NASEM (National Academies of Sciences, Engineering, and Medicine) (2018). *Sexual Harassment of Women: Climate, Culture, and Consequences in Academic Sciences, Engineering, and Medicine.* Washington, DC: The National Academies Press. https://doi.org/10.17226/24994.

National Academies of Sciences, Engineering, and Medicine. (2019a). *Minority Serving Institutions: America's Underutilized Resource for Strengthening the STEM Workforce.* Washington, DC: The National Academies Press. https://doi.org/10.17226/25257.

NASEM (2019b). *Taking Action Against Clinician Burnout: A Systems Approach to Professional Well-Being.* Washington, DC: The National Academies Press. https://doi.org/10.17226/25521.

NASEM (2019c). *The Science of Effective Mentorship in STEMM.* Washington, DC: The National Academies Press. https://www.nap.edu/read/25568.

NASEM (2020). *Promising Practices for Addressing the Underrepresentation of Women in Science, Engineering, and Medicine: Opening Doors.* Washington, DC: The National Academies Press. https://doi.org/10.17226/25585.

National Alliance for Caregiving (2019). Burning the candle at both ends: Sandwich generation caregiving in the U.S. https://www.caregiving.org/wp-content/uploads/2020/05/NAC-CAG_SandwichCaregiving_Report_Digital-Nov-26-2019.pdf.

National Alliance for Caregiving and AARP Public Policy Institute (2020). Caregiving in the U.S. 2020. https://www.aarp.org/content/dam/aarp/ppi/2020/05/full-report-caregiving-in-the-united-states.doi.10.26419-2Fppi.00103.001.pdf.

Newman, M. E. J. (2000). The structure of scientific collaboration networks. *Proceedings of the National Academy of Sciences of the United States of America*, 98(2), 404–409.

NIH (National Institutes of Health) (2020a). NIH Equity Committee. https://diversity.nih.gov/programs-partnerships/nih-equity-committee.

NIH (2020b). Flexibilities Available to Applicants and Recipients of Federal Financial Assistance Affected by COVID-19. Notice Number: NOT-OD-20-086. March 12, 2020. https://grants.nih.gov/grants/guide/notice-files/NOT-OD-20-086.html.

Niner, H. J., Johri, S., Meyer, J., and Wassermann, S. N. (2020). The pandemic push: Can COVID-19 reinvent conferences to models rooted in sustainability, equitability and inclusion? *Socio-Ecological Practice Research*, 2, 253–256. https://doi.org/10.1007/s42532-020-00059-y.

NSF (National Science Foundation) (2020). ADVANCE Program. *ADVANCE: Organizational Change for Gender Equity in STEM Academic Professions (ADVANCE).* https://www.nsf.gov/funding/pgm_summ.jsp?pims_id=5383.

NSPE (National Society of Professional Engineers) (2020). Open Forums in NSPE's Communities. https://www.nspe.org/resources/coronavirus-COVID-19-resources.

Nzinga, S. (2020). *Lean Semesters: How Higher Education Reproduces Inequity.* Baltimore, MD: Johns Hopkins University Press.

O'Brien, K. E., Biga, A., Kessler, S. R., and Allen, T. D. (2010). A meta-analytic investigation of gender differences in mentoring. *Journal of Management*, 36(2), 537–554. https://doi.org/10.1177/0149206308318619.

O'Connor, P., O'Hagan, C., Myers, E. S., Baisner, L., Apostolov, G., Topuzova, I., Saglamer, G., Tan, M. G., and Caglayan, H. (2020). Mentoring and sponsorship in higher education institutions: Men's invisible advantage in STEM? *Higher Education Research & Development*, 39(4), 764–777. https://doi.org/10.1080/07294360.2019.1686468.

OCUFA (Ontario Confederation of University Faculty Associations) (2020). Working Together to Build a Vibrant Future for Ontario's Universities. The Ontario Confederation of University Faculty Associations, https://ocufa.on.ca/blog-posts/working-together-to-build-a-vibrant-future-for-ontarios-universities/.

REFERENCES

OECD (Organisation for Economic Co-operation and Development) (2020). What is the impact of the COVID-19 pandemic on immigrants and their children? OECD: Policy Responses to Coronavirus (COVID-19). http://www.oecd.org/coronavirus/policy-responses/what-is-the-impact-of-the-COVID-19-pandemic-on-immigrants-and-their-children-e7cbb7de/#section-d1e105.

Oleschuk, M. (2020). Gender equity considerations for tenure and promotion during COVID 19. *Canadian Review of Sociology*, August 11, 2020. https://doi /10.1111/cars.12295.

Oliveira, D. F. M., Ma, Y., Woodruff, T. K., and Uzzi, B. (2019). Comparison of National Institutes of Health grant amounts to first-time male and female principal investigators. *JAMA, 321*(9), 898–900. https://doi.org/10.1001/jama.2018.21944.

Olson-Buchanan, J. B., and Boswell, W. R. (2006). Blurring boundaries: Correlates of integration and segmentation between work and nonwork. *Journal of Vocational Behavior, 68*(3), 432–445.

Ong, M., Wright, C., Espinosa, L., and Orfield, G. (2011). Inside the double bind: A synthesis of empirical research on undergraduate and graduate women of color in science, technology, engineering, and mathematics. *Harvard Educational Review, 81*(2), 172–209.

Oreopoulos, P., von Wachter, T., and Heisz, A. (2012). The short- and long-term career effects of graduating in a recession: Hysteresis and heterogeneity in the market for college graduates. *American Economic Journal: Applied Economics, 4*(1), 1–29. https://doi.org/10.1257/app.4.1.1.

Ornek, O. K., and Esin, M. N. (2020). Effects of a work-related stress model based mental health promotion program on job stress, stress reactions and coping profiles of women workers: A control groups study. *BMC Public Health, 20*(1), 1658. https://doi.org/10.1186/s12889-020-09769-0. PMID: 33148247; PMCID: PMC7641806.

Page, S. E. (2007). Making the difference: Applying a logic of diversity. *Academy of Management Perspectives, 21*(4), 6–20.

Page, S. E. (2008). *The Difference: How the Power of Diversity Creates Better Groups, Firms, Schools, and Societies*. Princeton, NJ: Princeton University Press.

Page, S. E. (2019). *The Diversity Bonus: How Great Teams Pay Off In The Knowledge Economy*. Princeton, NJ: Princeton University Press.

Pager, D., and Shepherd, H. (2008). The sociology of discrimination: Racial discrimination in employment, housing, credit, and consumer markets. *Annual Review of Sociology, 34*(1), 181–209.

Panagioti, M., Geraghty, K., Johnson, J., Zhou, A., Panagopoulou, E., Chew-Graham, C., Peters, D., Hodkinson, A., Riley, R., and Esmail, A. (2018). Association between physician burnout and patient safety, professionalism, and patient satisfaction. *JAMA Internal Medicine, 178*(10), 1317. https://doi.org/10.1001/jamainternmed.2018.3713.

Pappa, S., Ntella, V., Giannakas, T., Giannakoulis, V. G., Papoutsi, E., and Katsaounou, P. (2020). Prevalence of depression, anxiety, and insomnia among healthcare workers during the COVID-19 pandemic: A systematic review and meta-analysis. *Brain, Behavior, and Immunity, 88*, 901–907. https://doi.org/10.1016/j.bbi.2020.05.026. Epub May 8, 2020.

Park, C. L., Russell, B. S., Fendrich, M., Finkelstein-Fox, L., Hutchison, M., and Becker, J. (2020). Americans' COVID-19 Stress, Coping, and Adherence to CDC guidelines. *Journal of General Internal Medicine, 35*(8), 2296–2303. DOI:10.1007/s11606-020-05898-9.

Patton, E. W., Griffith, K. A., Jones, R. D., Stewart, A., Ubel, P. A., and Jagsi, R. (2017). Differences in mentor-mentee sponsorship in male vs female recipients of National Institutes of Health grants. *JAMA Internal Medicine*, published online February 20, 2017. https://doi.org/10.1001/jamainternmed.2016.9391.

Pearce, C. L., and Conger, J. A. (2003). *Shared Leadership: Reframing the How's and Why's of Shared Leadership*. Thousand Oaks, CA: Sage Publications.

Pearce, C. L., and Sims, Jr, H. P. (2002). Vertical versus shared leadership as predictors of the effectiveness of change management teams: An examination of aversive, directive, transactional, transformational, and empowering leader behaviors. *Group Dynamics: Theory, Research, and Practice, 6*(2), 172.

Pearce, C. L., Yoo, Y., and Alavi, M. (2004). Leadership, social work and virtual teams: The relative influence of vertical vs. shared leadership in the nonprofit sector. In R. E. Riggio and S. Smith-Orr (eds.), *Improving Leadership in Nonprofit Organizations*, 180–203. San Francisco, CA: Jossey-Bass.

Pérez, E. (2020). "'People of Color' Are Protesting. Here's What You Need to Know about This New Identity." *Washington Post*, July 2, 2020. https://www.washingtonpost.com/politics/2020/07/02/people-color-are-protesting-heres-what-you-need-know-about-this-new-identity/.

Pettit, E. (2020a). Being a woman in academe has its challenges. A global pandemic? Not helping. *Chronicle of Higher Education*, May 26, 2020. https://www.chronicle.com/article/being-a-woman-in-academe-has-its-challenges-a-global-pandemic-not-helping.

Pettit, E. (2020b). COVID-19 cuts hit contingent faculty hard. As the pandemic drags on, some question their future. *Chronicle of Higher Education*, October 26, 2020. https://www.chonicle.com/article/COVID-19-cuts-hit-contingent-faculty-hard-as-it-drags-on-some-question-their-future.

Phillips, K. W., Dumas, T. L., and Rothbard, N. P. (2018). Diversity and authenticity. *Harvard Business Review*, 96(2), 132–136.

Phillips, K. W., Rothbard, N. P., and Dumas, T. L. (2009). To disclose or not to disclose? Status distance and self-disclosure in diverse environments. *Academy of Management Review*, 34(4), 710–732. https://doi.org/10.5465/amr.34.4.zok710.

Pichevin, M., and Hurtig, M. (2007). On the necessity of distinguishing between sex and gender. *Feminism Psychology*, 17(4), 447–452. https://doi.org/10.1177/0959353507084324.

Pololi, L. H. (2010). *Changing the Culture of Academic Medicine: Perspectives of Women Faculty*. Lebanon, NH: University Press of New England.

Porter, E. (2017). "How care for elders denies women a paycheck." *The New York Times*, December 19, 2017. https://www.nytimes.com/2017/12/19/business/economy/women-work-elder-care.html.

Porter, K. B., Posselt, J. R., Reyes, K., Slay, K. E., and Kamimura, A. (2018). Burdens and benefits of diversity work: Emotion management in STEM doctoral students. *Studies in Graduate and Postdoctoral Education*, 9(2).

Porter, S., Toutkoushian, R., and Moore, J. (2008). Pay inequities for recently hired faculty, 1988-2004. *Review of Higher Education*, 31(4), 465–487. https://doi.org/31.. 10.1353/rhe.0.0014.

Powell, G. N., and Greenhaus, J. H. (2010). Sex, gender, and the work-to-family interface: Exploring negative and positive interdependencies. *Academy of Management Journal*, 53(3), 513–534.

Power, K. (2020). The COVID-19 pandemic has increased the care burden of women and families. *Sustainability: Science, Practice and Policy*, 16(1), 67–73.

Preis, H., Mahaffey, B., Heiselman, C., and Lobel, M. (2020a). Pandemic-related pregnancy stress and anxiety among women pregnant during the coronavirus disease 2019 pandemic. *American Journal of Obstetrics & Gynecology MFM*, 2(3), 100155. DOI:10.1016/j.ajogmf.2020.100155.

Preis, H., Mahaffey, B., Heiselman, C., and Lobel, M. (2020b). Vulnerability and resilience to pandemic-related stress among U.S. women pregnant at the start of the COVID-19 pandemic. *Social Science & Medicine*, 266, 113348. https://doi.org/10.1016/j.socscimed.2020.113348.

Pribbenow, C. M., Sheridan, J., Winchell, J., Benting, D., Handelsman, J., and Carnes, M. (2010). The tenure process and extending the tenure clock: The experience of faculty at one college. *Higher Education Policy*, 23, 17–38.

Prokos, A., and Padavic, I. (2002). "There oughtta be a law against bitches": Masculinity lessons in police academy training. *Gender, Work & Organization*, 9(4), 439–459. https://doi.org/10.1111/1468-0432.00168.

Purvanova, R. K., and Muros, J. P. (2010). Gender differences in burnout: A meta-analysis. *Journal of Vocational Behavior*, 77, 168–185. https://doi.org/10.1016/j.jvb.2010.04.006.

Que, J., Shi, L., Deng, J., Liu, J., Zhang, L., Wu, S., Gong, Y., Huang, W., Yuan, K., Yan, W., Sun, Y., Ran, M., Bao, Y., and Lu, L. (2020). Psychological impact of the COVID-19 pandemic on healthcare workers: A cross-sectional study in China. *Geneneral Psychiatry*, 33(3), e100259. https://doi.org/10.1136/gpsych-2020-100259. PMID: 32596640; PMCID: PMC7299004.

REFERENCES

Rabatin, J., Williams, E., Baier Manwell, L., Schwartz, M. D., Brown, R. L., and Linzer, M. (2016). Predictors and outcomes of burnout in Primary care physicians. *Journal of Primary Care & Community Health, 7*(1), 41–43.

Radecki, J., and Schonfeld, R. C. (2020). *The Impacts of COVID-19 on the Research Enterprise: A Landscape Review* (Research Report). New York: ITHAKA S+R. https://doi.org/10.18665/sr.214247.

Ragins, B. R., and McFarlin, D. B. (1990). Perceptions of mentor roles in cross-gender mentoring relationships. *Journal of Vocational Behavior, 37*(3), 321–339.

Raj, A., Carr, P. L., Kaplan, S. E., Terrin, N., Breeze, J. L., and Freund, K. M. (2016). Longitudinal analysis of gender differences in academic productivity among medical faculty across 24 medical schools in the United States. *Academic Medicine, 91*(8), 1074.

Raj, A., Kumra, T., Darmstadt, G. L., and Freund, K. M. (2019). Achieving gender and social equality: More than gender parity is needed. *Academic Medicine, 94*(11), 1658–1664. https://doi.org/10.1097/ACM.0000000000002877. PMID: 31335818.

Ramarajan, L., and Reid, E. (2020). Relational reconciliation: Socializing others across demographic differences. *Academy of Management Journal, 63*(2), 356–385. https://doi.org/10.5465/amj.2017.0506.

Redden, E. (2020). "COVID-19 roundup: Virtual falls and faculty pink slips." *Inside Higher Education*, July 21, 2020. https://www.insidehighered.com/news/2020/07/21/covid-roundup-colleges-revert-virtual-fall-canisius-and-carthage-plan-faculty.

Richards, A., Kanady, J. C., and Neylan, T. C. (2020). Sleep disturbance in PTSD and other anxiety-related disorders: An updated review of clinical features, physiological characteristics, and psychological and neurobiological mechanisms. *Neuropsychopharmacology, 45*(1), 55–73. https://doi.org/10.1038/s41386-019-0486-5. Epub August 23, 2019. Erratum in: *Neuropsychopharmacology*, October 7, 2019. PMID: 31443103; PMCID: PMC6879567.

Richards, A., Metzler, T. J., Ruoff, L. M., Inslicht, S. S., Rao, M., Talbot, L. S., and Neylan, T. C. (2013). Sex differences in objective measures of sleep in post-traumatic stress disorder and healthy control subjects. *Journal of Sleep Research, 22*(6), 679–687. https://doi.org/10.1111/jsr.12064. Epub June 14, 2013. PMID: 23763708; PMCID: PMC3958933.

Robbins, S., and Judge, T. A. (2018). *Organizational behavior (what's new in management)*. Harlow, England: Pearson.

Robertson, C., and Gebeloff, R. (2020). "How Millions of Women Became the Most Essential Workers in America." *New York Times*, April 18, 2020. https://www.nytimes.com/2020/04/18/us/coronavirus-women-essential-workers.html.

Rockquemore, K. A. (2015). Radical self care. Inside Higher Education website, May 6, 2015. https://www.insidehighered.com/advice/2015/05/06/essay-how-faculty-members-can-keep-focused-amid-so-much-disturbing-news.

Romero-Blanco, C., Rodríguez-Almagro, J., Onieva-Zafra, M. D., Parra-Fernández, M. L., Prado-Laguna, M. D. C., and Hernández-Martínez, A. (2020). Sleep pattern changes in nursing students during the COVID-19 lockdown. *International Journal of Environmental Research and Public Health, 17*(14), 5222.

Roper, R. L. (2019). Does gender bias still affect women in science? *Microbiology and Molecular Biology Review, 83*(3), e00018-19. DOI:10.1128/MMBR.00018-19.

Rothbard, N. P. (2001). Enriching or depleting? The dynamics of engagement in work and family roles. *Administrative Science Quarterly, 46*(4), 655–684. https://doi.org/10.2307/3094827.

Rothbard, N. P., Phillips, K. W., and Dumas, T. L. (2005). Managing multiple roles: Work-family policies and individuals' desires for segmentation. *Organization Science, 16*(3), 243–258. https://doi.org/10.1287/orsc.1050.0124.

Rothstein, J. (2020). The Lost Generation? Labor Market Outcomes for Post Great Recession Entrants. NBER Working Paper No. 27516, July 2020, JEL No. E24,J2. Cambridge, MA: National Bureau of Economic Research.

RSC (Royal Society of Chemistry) (2019). *Twitter Poster Conference.* http://www.rsc.org/events/detail/37540/rsc-twitter-poster-conference-2019.

Ruder, B., Plaza, D., Warner, R., and Bothwell, M. (2018). STEM women faculty struggling for recognition and advancement in a "men's club" culture. In *Exploring the Toxicity of Lateral Violence and Microaggressions,* 121–149. London, UK: Palgrave Macmillan, Cham.

Ruzycki, S. M., Fletcher, S., Earp, M., Bharwani, A, and Lithgow, K. C. (2019). Trends in the proportion of female speakers at medical conferences in the United States and in Canada, 2007 to 2017. *JAMA Network Open, 2*(4), e192103. https://doi.org/10.1001/jamanetworkopen.2019.2103.

Ryan, G. W., and Bernard, H. R. (2003). Techniques to identify themes. *Field Methods, 15*(1), 85–109.

Sandler, B. R., and Hall, R. M. (1986). The campus climate revisited: Chilly for women faculty, administrators, and graduate students. https://eric.ed.gov/?id=ED282462.

Santamaría, L. J. (2014). Critical change for the greater good: Multicultural perceptions in educational leadership toward social justice and equity. *Educational Administration Quarterly, 50*(3), 347–391.

Sarabipour, S., Schwessinger, B., Mumoki, F. N., Mwakilili, A. D., Khan, A., Debat, H. J., Sáez, P. J., Seah, S., and Mestrovic, T. (2020). Evaluating features of scientific conferences: A call for improvements. *bioRxiv,* 2020.04.02.022079. https://doi.org/10.1101/2020.04.02.022079.

Savage, M. (2020). As working mums perform more childcare and face increased job insecurity, there are fears COVID-19 has undone decades of advancement. But could the pandemic be a catalyst for progress? BBC: *Worklife.* https://www.bbc.com/worklife/article/20200630-how-COVID-19-is-changing-womens-lives.

Schiebinger, L., Henderson, A. D., and Gilmartin, S. K. (2008). *Dual Career Academic Couples: What Universities Need to Know.* Machelle R. Clayman Institute for Gender Research. Stanford, CA: Stanford University. https://gender.stanford.edu/sites/g/files/sbiybj5961/f/publications/dualcareerfinal_0.pdf.

Schiffer, P., and Walsch, J. (2020). Known unknowns. *Inside Higher Education,* October 13, 2020. https://www.insidehigher.com/views/2020/10/13/impact-social-distancing-process-and-outcomes-university-research-opinion.

Schmidt, G. (2020). "As Labs Reopen. USC Researchers Adjust to New Campus Guidelines." *USC News,* July 13, 2020. https://news.usc.edu/173160/usc-research-labs-reopen-campus-guidelines-COVID-19.

Schmitt, M. T., Branscombe, N. R., Postmes, T., and Garcia, A. (2014). The consequences of perceived discrimination for psychological wellbeing: A meta-analytic review. *Psychological Bulletin, 140,* 921–948. https://doi.org/10.1037/a0035754.

Schreiber, M. (2020). "Female scientists are bearing the brunt of quarantine child-rearing." *The New Republic,* May 22, 2020. https://newrepublic.com/article/157785/female-scientists-bearing-brunt-quarantine-child-rearing.

Schreier, M. (2012). Qualitative content analysis in practice. In U. Flick (ed.), *The SAGE Handbook of Qualitative Data Analysis,* 170–183. London, UK: Sage Publications.

Schwandt, H. (2019). Recession Graduates: The Long-lasting Effects of an Unlucky Draw. Stanford Institute for Economic Policy Research Policy Brief, April 2019. https://siepr.stanford.edu/sites/default/files/publications/PolicyBrief-Apr2019.pdf.

Scott, J. (2016). "Unintended Help for Male Professors." *Inside Higher Education,* June 27, 2016. https://www.insidehighered.com/news/2016/06/27/stopping-tenure-clock-may-help-male-professors-more-female-study-finds.

Segarra, V. A., Vega, L. R., Primus, C., Etson, C., Guillory, A. N., Edwards, A., Flores, S. C., Fry, C., Ingram, S. L., Lawson, M., McGee, R., Paxson, S., Phelan, L., Suggs, K., Vuong, E., Hammonds-Odie, L., Leibowitz, M. J., Zavala, M. E., Lujan, J. L., and Ramirez-Alvarado, M. (2020). Scientific societies fostering inclusive scientific environments through travel awards: Current practices and recommendations. *CBE—Life Sciences Education, 19*(2), 1–10. https://doi.org/10.1187/cbe.19-11-0262.

REFERENCES

Sevilla, A., and Smith, S. (2020). Baby Steps: The Gender Division of Childcare During the COVID-19 Pandemic. IZA Discussion Paper No. 13302. Bonn, DE: Institute of Labor Economics.

Shakeshaft, C., Brown, G., Irby, B. J., Grogan, M., and Ballenger, J. (2007). Increasing gender equity in educational leadership. In S. S. Klein (ed.), *Handbook for achieving Gender Equity through Education*, 103–129. Mahwah, NJ: Lawrence Erlbaum Associates.

Sharma, M. K., Anand, N., Singh, P., Vishwakarma, A., Mondal, I., Thakur, P. C., and Kohli, T. (2020). Researcher burnout: An overlooked aspect in mental health research in times of COVID-19. *Asian Journal of Psychiatry, 54*, 102367. https://doi.org/10.1016/j.ajp.2020.102367.

Shaukat, N., Ali, D. M., and Razzak, J. (2020). Physical and mental health impacts of COVID-19 on healthcare workers: A scoping review. *International Journal of Emergency Medicine, 13*(1), 40. https://doi.org/10.1186/s12245-020-00299-5.

Shaw, J., and Chew, Y. L. (2020). Early and mid-career scientists face a bleak future in the wake of the pandemic. The Conversation website, August 13, 2020. https://theconversation.com/early-and-mid-career-scientists-face-a-bleak-future-in-the-wake-of-the-pandemic-144350.

Sheraton, M., Deo, N., Dutt, T., Surani, S., Hall-Flavin, D., and Kashyap, R. (2020). Psychological effects of the COVID 19 pandemic on healthcare workers globally: A systematic review. *Psychiatry Research, 292*, 113360. https://doi.org/10.1016/j.psychres.2020.113360. Epub August 3, 2020. PMID: 32771837.

Shields, C. M. (2010). Transformative leadership: Working for equity in diverse contexts. *Educational Administration Quarterly, 46*(4), 558–589.

Shockley, K. M., Clark, M. A., Dodd, H., and King, E. B. (2020). Work-family strategies during COVID-19: Examining gender dynamics among dual-earner couples with young children. *Journal of Applied Psychology, 106*(1), 15–28. http://dx.doi.org/10.1037/apl0000857.

Shockley, K. M., and Shen, W. (2016). Couple dynamics: Division of labor. In T. Allen and L. Eby (eds.), *Oxford Handbook of Work and Family*. Oxford University Press.

Shockley, K. M., Shen, W., DeNunzio, M. M., Arvan, M. L., and Knudsen, E. A. (2017). Disentangling the relationship between gender and work–family conflict: An integration of theoretical perspectives using meta-analytic methods. *Journal of Applied Psychology, 102*(12), 1601–1635. https://doi.org/10.1037/apl0000246.

Sladek, M. R., Doane, L. D., Jewell, S. L., and Luecken, L. J. (2017). Social support coping style predicts women's cortisol in the laboratory and daily life: The moderating role of social attentional biases. *Anxiety, Stress & Coping, 30*(1), 66–81. https://doi.org/10.1080/10615806.2016.1181754. Epub May 18, 2016. PMID: 27189781.

Slaughter, S., and Rhoades, G. (2004). *Academic Capitalism and the New Economy: Markets, State, and Higher Education*. Baltimore, MD: Johns Hopkins University Press.

Smith, C. A. S. (2014). Assessing Academic STEM Women's Sense of Isolation in the Workplace. In P. J. Gilmer, B. Tansel, and M. H. Miller (eds.), *Alliances for Advancing Academic Women. Bold Visions in Educational Research*, 97–117. Rotterdam, NL: SensePublishers. https://doi.org/10.1007/978-94-6209-604-2_5.

Smith, K. (2019). Invisible work: A qualitative study of the emotional labor of professors. Ph.D. dissertation, Indiana University of Pennsylvania.

Somerville, L. H., and Gruber, J. (2020). "Three trouble spots facing women in science—and how we can tackle them." *Science*, October 16, 2020. https://www.sciencemag.org/careers/2020/10/three-trouble-spots-facing-women-science-and-how-we-can-tackle-them.

Southworth, C., Finn, J., Dawson, S., Fraser, C., and Tucker, S. (2007). Intimate partner violence, technology, and stalking. *Violence Against Women, 13*(8), 842–856. https://doi.org/10.1177/1077801207302045.

Spitzer, R. L., Kroenke, K., Williams, J. B. W., and Löwe, B. (2006). A brief measure for assessing generalized anxiety disorder: The GAD-7. *Archives of Internal Medicine, 166*(10), 1092–1097. https://doi.org/10.1001/archinte.166.10.1092.

Squazzoni, F., Bravo, G., Grimaldo, F., Garcia-Costa, D., Farjam, M., and Mehmani, B. (2020). No Tickets for Women in the COVID-19 Race? A Study on Manuscript Submissions and Reviews in 2347 Elsevier Journals during the Pandemic. Social Science Research Network, October 16, 2020. https://ssrn.com/abstract=3712813 or http://dx.doi.org/10.2139/ssrn.3712813.

Stanford University (2020). COVID-19 Tenure and Appointment Clock Extension Policy. *Faculty Handbook.* https://facultyhandbook.stanford.edu/COVID-19-tenure-and-appointment-clock-extension-policy.

Staniscuasky, F., Reichert, F., Werneck, F. P., de Oliveira, L., Mello-Carpes, P. B., Soletti, R. C., Almeida, C. I., Zandona, E., Ricachenevsky, F. K., Neumann, A., Schwartz, I. V. D., Tamajusuku, A. S. K., Seixas, A., Kmetzsch, L., and Parent in Science Movement. (2020). Impact of COVID-19 on academic mothers. *Science, 368*(6492), 724. doi:10.1126/science.abc2740.

Stanko, T. L., and Beckman, C. M. (2014). Watching you watching me: Boundary control and capturing attention in the context of ubiquitous technology use. *Academy of Management Journal, 58*(3), 712–738. https://doi.org/10.5465/amj.2012.0911.

Stelitano, L., Doan, S., Woo, A., Diliberti, M., Kaufman, J. H., and Henry, D. (2020). The Digital Divide and COVID-19: Teachers' Perceptions of Inequities in Students' Internet Access and Participation in Remote Learning. Santa Monica, CA: Rand Corp. https://www.rand.org/pubs/research_reports/RRA134-3.html.

Strong, E. A., De Castro, R., Sambuco, D., Stewart, A., Ubel, P. A., Griffith, K. A., and Jagsi, R. (2013). Work-life balance in academic medicine: Narratives of physician-researchers and their mentors. *Journal of General Internal Medicine, 28*(12), 1596–1603. https://doi.org/10.1007/s11606-013-2521-2. Epub June 14, 2013.

Sutter, M., and Perrin, P. B. (2016). Discrimination, mental health, and suicidal ideation among LGBTQ people of color. *Journal of Counseling Psychology, 63*(1), 98–105. https://doi.org/10.1037/cou0000126. PMID: 26751158.

SWE (Society of Women Engineers) (2020). *Impact of COVID-19 on Women in Engineering and Technology.* Survey Report, July 2020. Chicago, IL: Society of Women Engineers. https://swe.org/wp-content/uploads/2020/07/SWE-COVID-19-July-2020.pdf.

Taylor, A. (2020). Beyond the Neoliberal university. *Boston Review*, August 4, 2020. http://bostonreview.net/class-inequality/todd-wolfson-astra-taylor-beyond-neoliberal-university.

Taylor, B. J., and Cantwell, B. (2018). Unequal higher education in the United States: Growing participation and shrinking opportunities. *Social Sciences, 7*(9), 167.

Taylor, B. J., and Cantwell, B. (2019). *Unequal Higher Education: Wealth, Status, and Student Opportunity.* New Brunswick, NJ: Rutgers University Press.

Taylor Jr., H. L., Kwiatek, B., and Luter, G. (2020). *Centering Race and Anti-Racism: The University & the Post-COVID-19 World.* University of Pennsylvania. Philadelphia, PA: Netter Center for Community Partnerships. https://www.nettercenter.upenn.edu/sites/default/files/UCS_Journal_Volume10_Fall2020.pdf#page=114.

Taylor, M. P. (2019). Today's affinity groups: Risks and rewards. Society for Human Resource Management website, October 11, 2019. https://www.shrm.org/resourcesandtools/legal-and-compliance/employment-law/pages/affinity-groups-risks-rewards.aspx.

Taylor, S., Landry, C. A., Paluszek, M. M., Fergus, T. A., McKay, D., and Asmundson, G. J. G. (2020). COVID stress syndrome: Concept, structure, and correlates. *Depression and Anxiety, 37*(8), 706–714. https://doi.org/10.1002/da.23071. Epub July 5, 2020. PMID: 32627255; PMCID: PMC7362150.

Taylor, S.E., Klein, L.C., Lewis, B.P., Gruenewald, T.L., Gurung, R.A.R., Updegraff, J.A. (2000). Biobehavioral responses to stress in females: tend-and-befriend, not fight-or- flight. *Psychological review*, 107(3), 411–429. https://doi.org/10.1037/0033-295x.107.3.411.

Terosky, A. L., and Gonzales, L. D. (2016). Re-envisioned contributions: Experiences of faculty employed at institutional types that differ from their original aspirations. *Review of Higher Education, 39*(2), 241–268.

REFERENCES

Theoharis, G. (2007). Social justice educational leaders and resistance: Toward a theory of social justice leadership. *Educational Administration Quarterly*, *43*(2), 221–258.

Thompson, M. (2020). Balance is bunk: Organizational and marital turnover in dual academic career couples. In E. Kossek and K.-H. Lee (eds.), *Fostering Gender and Work-Life Inclusion for Faculty in Understudied Contexts: An Organizational Science Lens*, 116–120. West Lafayette, IN: Purdue e-Pubs. https://doi.org/10.5703/1288284317224. https://docs.lib.purdue.edu/worklifeinclusion/2018/ dccsowfs/3.

Trockel, M., Bohman, B., Lesure, E., Hamidi, M. S., Welle, D., Roberts, L., and Shanafelt, T. (2018). A brief instrument to assess both burnout and professional fulfillment in physicians: Reliability and validity, including correlation with self-reported medical errors, in a sample of resident and practicing physicians. *Academic Psychiatry*, *42*(1), 11–24. https://doi.org/10.1007/s40596-017-0849-3. Epub December 1, 2017. PMID: 29196982; PMCID: PMC5794850.

Tunguz, S. (2016). In the eye of the beholder: Emotional labor in academia varies with tenure and gender. *Studies in Higher Education*, *41*(1), 3–20.

Turner, C.S.V. and González, J.C., (2011). Faculty women of color: The critical nexus of race and gender. *Journal of Diversity in Higher Education*, *4*(4), 199.

UCB (University of California, Berkeley) (2020). "COVID-19 impacts Berkeley budget, reduction measures required." July 15, 2020. https://news.berkely.edu/2020/07/15/update-on-budget-expense-reduction-measures.

UCL (University College London) (2020). COVID-19-impact on research funding. UCL website, September 2, 2020. https://www.ucl.ac.uk/research-services/news/2020/sep/COVID-19-impact-research-funding.

UCLA (University of California, Los Angeles) (2020). Center for the Study of Women. Open letter to chancellor on research productivity and childcare, July 6, 2020. https://csw.ucla.edu/2020/07/06/open-letter-on-research-productivity-and-childcare/.

Umberson, D., and Montez, J. K. (2010). Social relationships and health: A flashpoint for health policy. *Journal of Health and Social Behavior*, *51*(Suppl), S54–S66. https://doi.org/10.1177/0022146510383501. PMID: 20943583; PMCID: PMC3150158.

UMR (University of Minnesota, Rochester) (2020). "Responding to the financial challenges of the COVID-19 pandemic." April 15, 2020. https://www.rochester.edu/coronavirus-updates/responding-to-the-financial-challenges-of-the-COVID-19-pandemic.

UN (United Nations) (2020). Policy brief: The impact of COVID-19 on women. April 9, 2020. https://www.un.org/sexualviolenceinconflict/wp-content/uploads/2020/06/report/policy-brief-the-impact-of-COVID-19-on-women/policy-brief-the-impact-of-COVID-19-on-women-en-1.pdf.

University of Toronto, Division of the Vice-President & Provost (2020). UTogether: A Roadmap for the University of Toronto. https://www.provost.utoronto.ca/planning-policy/utogether2020-a-roadmap-for-the-university-of-toronto/.

U.S. Census Bureau (2020). "U.S. Department of Commerce Secretary Wilbur Ross and U.S. Census Director Steven Dillingham Statement on 2020 Census Operational Adjustments Due to COVID-19." Release No. CV20-RTQ.16, April 13, 2020. https://www.census.gov/newsroom/press-releases/2020/statement-COVID-19-2020.html.

UT System (University of Texas System), Board of Regents (2020). "Probationary Period Extensions-Under COVID-19." April 16, 2020. https://provost.utexas.edu/the-office/faculty-affairs/probationary-period-extensions-for-tenure-track-faculty.

UW (University of Washington), Office of Academic Personnel (2020). Promotion/Tenure Clock Extensions Due to COVID-19 – Faculty. https://ap.washington.edu/ahr/working/promotion-and-tenure-extensions/extension-of-promotion-tenure-clock-due-to-COVID-19/.

UW System (University of Wisconsin System) (2020). Women's and Gender Studies Consortium. *Statement Regarding Caregiving Recommendations*. Caregiving Task Force, June 23, 2020. https://consortium.gws.wisc.edu/caregiving-task-force/.

Vaghela, P., and Sutin, A. R. (2016). Discrimination and sleep quality among older US adults: The mediating role of psychological distress. *Sleep Health*, 2(2), 100–108. https://doi.org/10.1016/j.sleh.2016.02.003. Epub April 4, 2016. PMID: 28923251.

Vanderbilt, A. A., Isringhausen, K. T., VanderWielen, L. M., Wright, M. S., Slashcheva, L. D., and Madden, M, A. (2013). Health disparities among highly vulnerable populations in the United States: A call to action for medical and oral health care. *Medical Education Online*, 18, 1–3. https://doi.org/10.3402/meo.v18i0.20644. PMID: 23534859; PMCID: PMC3609999.

van der Woude, D., and van der Helm-van Mil, A. H. M. (2018). Update on the epidemiology, risk factors, and disease outcomes of rheumatoid arthritis. *Best Practice & Research: Clinical Rheumatology*, 32(2),174–187.

van Veen, R., and Wijnants, R. (2020). Division of work between fathers and mothers is changed by the corona crisis. Utrecht University website, *News*, May 25, 2020. https://www.uu.nl/en/news/division-of-work-between-fathers-and-mothers-is-changed-by-the-corona-crisis.

Veijola, S., and Jokinen, E. (2018). Coding gender in academic capitalism. *Ephemera: Theory and Politics in Organization*, 18(3), 527–549.

Viglione, G. (2020). Are women publishing less during the pandemic? Here's what the data say. *Nature*, 581, 365–366. https://media.nature.com/original/magazine-assets/d41586-020-01294-9/d41586-020-01294-9.pdf.

Vincent-Lamarre, P., Sugimoto C.R. and Larivière V. (2020). The Decline of Women's Research Production during the Coronavirus Pandemic. *Nature Index*. May 19, 2020. https://www.natureindex.com/news-blog/decline-women-scientist-research-publishing-production-coronavirus-pandemic.

Vogels, E.A., Perrin, A., Rainie, L. and Anderson, M. (2020). "53% of Americans Say the Internet Has Been Essential During the COVID-19 Outbreak." On Pew Research Center Website, April 30, 2020. https://www.pewresearch.org/internet/2020/04/30/53-of-americans-say-the-internet-has-been-essential-during-the-COVID-19-outbreak.

Voltmer, E., Obst, K., and Kötter, T. (2019). Study-related behavior patterns of medical students compared to students of science, technology, engineering and mathematics (STEM): A three-year longitudinal study. *BMC Medical Education*, 19(1), 262. https://doi.org/10.1186/s12909-019-1696-6. PMID: 31307437; PMCID: PMC6631808.

von Känel, R., Princip, M., Holzgang, S. A., Fuchs, W. J., van Nuffel, M., Pazhenkottil, A. P., and Spiller, T. R. (2020). Relationship between job burnout and somatic diseases: A network analysis. *Scientific Reports*, 10(1), 18438. https://doi.org/10.1038/s41598-020-75611-7. PMID: 33116176; PMCID: PMC7595180.

Voosen, P. (2020). Coronavirus forces United States, United Kingdom to cancel Antarctic field research. *Science*, June 12, 2020. https://doi.org/10.1126/science.abd3057.

Wallheimer, B. (2020). Non-tenured and female faculty feeling COVID burdens, study says. Purdue University, December 15, 2020. https://www.purdue.edu/newsroom/releases/2020/Q4/non-tenured-and-female-faculty-feeling-covid-burdens,-study-says.html.

Ward, K., and Wolf-Wendel, L. (2012). *Academic motherhood: How faculty manage work and family*. New Brunswick, NJ: Rutgers University Press.

Way, S. F., Morgan, A. C., Larremore, D. B., and Clauset, A. (2019). Productivity, prominence, and the effects of academic environment. *Proceedings of the National Academy of Sciences of the United States of America*, 116(22), 10729–10733.

Weber, F. C., Norra, C., and Wetter, T. C. (2020). Sleep disturbances and suicidality in posttraumatic stress disorder: An overview of the literature. *Frontiers in Psychiatry*, 11, 167. https://doi.org/10.3389/fpsyt.2020.00167. PMID: 32210854; PMCID: PMC7076084.

Wheatley, M. J. (1999). *Leadership and the New Science: Discovering Order in a Chaotic World* (2nd ed.). Oakland, CA: Berrett-Koehler.

Whitlock, J. (2020). COVID-19 stalls clinical trials for everything but COVID-19. *WIRED*, April 17, 2020. https://www.wired.com/story/COVID-19-stalls-clinical-trials-for-everything-but-COVID-19/.

REFERENCES

WHO (World Health Organization) (2019). "QD85: Burn-out." Departmental News, May 28, 2019. https://www.who.int/news/item/28-05-2019-burn-out-an-occupational-phenomenon-international-classification-of-diseases.

WHO (2021). Health Systems: Equity. WHO website, Health Systems Topics page. https://www.who.int/healthsystems/topics/equity/en/.

Wilkerson, A. (2020). Student to scholar: Critical ethnographic conceptualizations of mentoring a black female scholar and considerations for diversifying the academy. In D. Chapman and S. Wilkerson (eds.), *From Student to Scholar*, 1–17. https://doi.org/10.1007/978-3-030-42081-9.

Williams, D. R. (2018). Stress and the mental health of populations of color: Advancing our understanding of race-related stressors. *Journal of Health and Social Behavior*, 59(4), 466–485. https://doi.org/10.1177/0022146518814251. PMID: 30484715; PMCID: PMC6532404.

Williams, J. C. (2020). The pandemic has exposed the fallacy of the "ideal worker." *Harvard Busines Review*, May 11, 2020.

Woitowich, N. C., Jain, S., Arora, V. M., and Joffe, H. (2020). COVID-19 threatens progress toward gender equity within academic medicine. *Academic Medicine*, September 29, 2020. https://doi.org/10.1097/ACM.0000000000003782. Epub ahead of print.

Wolfinger, N. H., Mason, M. A., and Goulden, M. (2008). Problems in the pipeline: Gender, marriage, and fertility in the ivory tower. *Journal of Higher Education*, 79(4), 388–405.

Wooden, P., and Hanson, B. (2020). Earth and space science collaboration during the COVID-19 pandemic: An analysis of AGU journal and fall meeting authors and teams, 2018–2020. Poster submission to the American Geophysical Union. http://agu2020fallmeeting-agu.ipostersessions.com/Default.aspx?s=56-A6-F5-19-4B-C4-D0-EE-3A-5B-77-59-0D-C3-C6-7D.

Woolston, C. (2019). PhDs: The tortuous truth. *Nature*, 575, 403–406. doi: https://doi.org/10.1038/d41586-019-03459-7.

Woolston, C. (2020a). Seeing an 'exit plan' for leaving academia amid coronavirus worries. *Nature*, July 6, 2020. https://www.nature.com/articles/d41586-020-02029-6.

Woolston, C. (2020b). It's like we're going back 30 years: How the coronavirus is gutting diversity in science. *Nature*, July 31, 2020. https://doi.org/10.1038/d41586-020-02288-3.

Woolston, C. (2020c). Pandemic darkens postdocs' work and career hopes. *Nature*, September 8, 2020. https://www.nature.com/articles/d41586-020-02548-2?sf237612051=1.

Woolston, C. (2020d). Signs of depression and anxiety soar among US graduate students during pandemic. *Nature*, 585(7823), 147–148. https://doi.org/10.1038/d41586-020-02439-6. PMID: 32811983.

Woolston, C. (2020e). Postdocs under pressure: "Can I even do this anymore?" *Nature*, 587(7835), 689–692. https://doi.org/10.1038/d41586-020-03235-y. PMID: 33230311.

Wotschack, P., Glebbeek, A., and Wittek, R. (2014). Strong boundary control, weak boundary control and tailor-made solutions: The role of household governance structures in work–family time allocation and mismatch. *Community, Work & Family*, 17(4), 436–455. https://doi.org/10.1080/13668803.2014.923380.

Wyatt, J., and Robertson, N. (2011). Burnout in university teaching staff: A systematic literature review. *Educational Research*, 53(1), 33050.

Yank, V., Rennels, C., Linos, E., Choo, E. K., Jagsi, R., and Mangurian, C. (2019). Behavioral health and burnout among physician mothers who care for a person with a serious health problem, long-term illness, or disability. *JAMA Internal Medicine*, 179(4), 571–574. https://doi.org/10.1001/jamainternmed.2018.6411.

Yaribeygi, H., Panahi, Y., Sahraei, H., Johnston, T. P., and Sahebkar, A. (2017). The impact of stress on body function: A review. *EXCLI Journal*, 16, 1057–1072. https://doi.org/10.17179/excli2017-480. PMID: 28900385; PMCID: PMC5579396.

Yavorsky, J. E., Dush, C. M., and Schoppe-Sullivan, S. J. (2015). The production of inequality: The gender division of labor across the transition to parenthood. *Journal of Marriage and the Family*, 77(3), 662–679. https://doi.org/10.1111/jomf.12189.

Yeager, A. (2020). How the COVID-19 pandemic has affected field research. *Scientist*, August 20, 2020. https://www.the-scientist.com/news-opinion/how-the-COVID-19-pandemic-has-affected-field-research-67841.

Yoshino, K. (2001). Covering. *Yale Law Journal*, *11*(4), 769–939. https://doi.org/10.2307/797566.

Yu, L., Buysse, D. J., Germain, A., Moul, D. E., Stover, A., Dodds, N. E., Johnston, K. L., and Pilkonis, P. A. (2011). Development of short forms from the PROMIS™ sleep disturbance and sleep-related impairment item banks. *Behavioral Sleep Medicine*, *10*(1), 6–24. https://doi.org/10.1080/15402002.2012.636266.

Zahneis, M. (2020). For many graduate students, COVID-19 pandemic highlights inequities. *Chronicle of Higher Education*, March 26, 2020. https://www.chronicle.com/article/for-many-graduate-students-COVID-19-pandemic-highlights-inequities/.

Zamarro, G., and Prado, M. J. (2020). Gender Differences in Couples' Division of Childcare, Work and Mental Health during COVID-19. CESR-Schaeffer Working Paper No. 003, August 8, 2020. https://papers.ssrn.com/sol3/papers.cfm?abstract_id=3667803.

Zeng, L. N., Yang, Y., Wang, C., Li, X. H., Xiang, Y. F., Hall, B. J., Ungvari, G. S., Li, C. Y., Chen, C., Chen, L. G., Cui, X. L., An, F. R., and Xiang, Y. T. (2019). Prevalence of poor sleep quality in nursing staff: A meta-analysis of observational studies. *Behavioral Sleep Medicine*, *18*(6), 746–759. https://doi.org/10.1080/15402002.2019.1677233. Epub October 31, 2019. PMID: 31672062.

Zimmer, K. (2020). Gender gap in research output widens during pandemic. *Scientist*, June 25, 2020. https://www.the-scientist.com/news-opinion/gender-gap-in-research-output-widens-during-pandemic-67665.

Zimmerman, C. A., Carter-Sowell, A. R., and Xu, X. (2016). Examining workplace ostracism experiences in academia: Understanding how differences in the faculty ranks influence inclusive climates on campus. *Frontiers in Psychology*, *7*, 753.

Zreik, G., Asraf, K., Haimov, I., and Tikotzky, L. (2020). Maternal perceptions of sleep problems among children and mothers during the coronavirus disease 2019 (COVID-19) pandemic in Israel. *Journal of Sleep Research*, September 29, 2020, e13201. https://doi.org/10.1111/jsr.13201. Epub ahead of print. PMID: 32996188; PMCID: PMC7536915.

Zschirnt, E. (2016). Measuring Hiring Discrimination – A History of Field Experiments in Discrimination Research. NCCR–On the Move, Working Paper Series No. 7, May 16, 2016. http://dx.doi.org/10.2139/ssrn.2780398.

Appendix A

Literature Review Terms and Survey Methodology for "Boundaryless Work: The Impact of COVID-19 on Work-Life Boundary Management, Integration, and Gendered Divisions of Labor for Academic Women in STEMM," by Ellen Ernst Kossek, Tammy D. Allen, and Tracy L. Dumas

SURVEY METHODOLOGY

We[1] designed a survey to ask women faculty in science, technology, engineering, mathematics, and medicine (STEMM) to compare how the COVID-19 pandemic has affected them between March 2020 and October 2020. Using a mixed-methods approach including qualitative and quantitative formats, the survey asked respondents to indicate their work-location preferences and boundary control; changes in work-life coping strategies, childcare and eldercare, and other domestic demands; and preferences for university support. The survey was publicized on the ADVANCE grant listserv and listservs of academic women in scientific specialties (see Table A-1).

We presented the results of 933 faculty who identified themselves as STEMM faculty and provided usable data. The final sample included 763 respondents; other respondents who were omitted were not women (25), not in STEMM (190), did not indicate STEMM status (286), or had other incomplete data. We focused our report on the results from 733 tenured or tenure-stream faculty, since these individuals, in addition to teaching and service roles, were juggling research demands that may have results in significant career setbacks that could harm tenure, research funding and implementation, and promotion. We have also included some data from the 170 non-tenure-stream respondents. Many of their concerns mirrored those of tenure-stream faculty. Table A-2 shows sample demographic breakouts.

---

[1] In this appendix, Kossek, Allen, and Dumas use the first person plural.

TABLE A-1 Listservs that Posted the Anonymous Survey Link for the Work-Life Boundaries Paper

American Society for Engineering Education (ASEE) list of Engineering Deans
Big Ten+ Associate Deans of Engineering for Academic Affairs
ASEE Women in Engineering Division
Women in Engineering Pro-Active Network
Computing Research Association Committee on Widening Participation
Association for Computing Machinery (ACM) website
ACM Council on Women in Computing
National Academy of Engineering list of women members
Purdue Women Faculty in Engineering (1) mailing list; (2) Dean of Engineering Thursday Memo; and (3) Deans of the Colleges of (a) Agriculture, (b) Pharmacy, (c) Purdue Polytechnic Institute, and (d) Veterinary Medicine & Purdue American Association of University Professor Twitter

Survey Sample Data and Analytical Approach

Nearly all (98 percent) of the 763 tenure-track or tenured women faculty in STEMM fields were from 202 U.S. institutions, and a small number (a little more than 2 percent or n = 20) of participants were from non-U.S. institutions. The survey was distributed on U.S. listservs. About half the respondents, or 326 people, were from 77 R1 institutions. The sample had representation from many disciplines as follows: industrial, material, and general engineering (n = 129, 16.9 percent); chemistry, chemical engineering, biology, and biochemistry (n = 102, 13.9 percent); health sciences (n = 56, 7.3 percent); electrical and mechanical engineering (n = 48, 6.3 percent); mathematics and statistics (n = 27, 3.5 percent); atmospheric, Earth, and ocean sciences (n = 25, 3.3 percent); agriculture and natural resources (n = 17, 2.2 percent); physics (n = 9, 1.2 percent); and other disciplines. For rank, the sample was evenly distributed with about one-third (34.1 percent) untenured assistant professors, one-third associate professors (31.2 percent), and one-third (34.7 percent) full professors. Approximately three-fourths of the sample was white (72.9 percent) and married or living with a romantic partner (86.5 percent). A little less than one-tenth (7.3 percent) of married women faculty lived apart from their spouse or one of the spouses lived far from work because of the other's work. More than half (58.2 percent) provided care for children under the age of 18, 10.4 percent provided eldercare, 3.9 percent provided sandwiched care (i.e., both child care and eldercare). Nearly one-fifth, or 17.8 percent, provided care for family members who do not live with them.

The sample of women faculty in STEMM fields who are not on the tenure track included 170 participants predominantly from 62 U.S. institutions. The survey population was composed of faculty (91.2 percent), researchers (5.9 percent),

TABLE A-2 Sample Description for October 2020 Survey of Women in Academic STEMM Faculty

Characteristics		Tenured or on the Tenure Track (n = 763) N (%)				Non-Tenure-Track (n = 170) N (%)				Study Sample	
		Assistant (n = 258, 34.1)	Associate (n = 236, 31.2)	Full (n = 263, 34.7)	Total (n = 763)	Faculty (n = 155, 91.2)	Researcher (n = 10, 5.9)	Postdocs (n = 5, 2.9)	Total (n = 170)		Total (n = 933)
Ethnicity	White	172 (66.7)	171 (72.5)	209 (88.9)	556 (72.9)	116 (74.8)	10 (100)	2 (40.0)	128 (75.3)		684 (73.3)
	Hispanic	20 (7.8)	17 (7.2)	13 (4.9)	50 (6.6)	12 (7.7)	0	1 (20.0)	13 (7.6)		63 (6.8)
	Black	6 (2.3)	4 (1.7)	1 (0.4)	11 (1.4)	7 (4.5)	0	0	7 (4.1)		18 (109)
	Asian/Pacific Is.	31 (12.0)	23 (9.7)	16 (6.1)	70 (9.2)	7 (4.5)	0	1 (20.0)	8 (4.7)		78 (8.4)
	Multi-Racial/Other	11 (4.3)	8 (3.4)	9 (3.4)	28 (3.7)	7 (4.5)	0	1 (20.0)	8 (4.7)		36 (3.9)
Relationship	Married	198 (76.7)	183 (77.5)	215 (81.7)	601 (80.2)	120 (77.4)	7 (70.0)	3 (60.0)	130 (76.5)		731 (78.3)
	Living with a Romantic Partner	23 (8.9)	16 (6.8)	9 (3.4)	49 (6.3)	9 (5.8)	0	1 (20.0)	10 (5.9)		59 (6.3)
	Single	28 (10.9)	34 (14.4)	33 (12.5)	95 (12.9)	24 (15.5)	3 (30.0)	1 (20.0)	28 (16.5)		123 (13.2)
	Widowed	1 (0.4)	1 (0.4)	3 (1.1)	5 (0.7)	1 (0.6)	0	0	1 (0.6)		6 (0.6)
	Long-Distance Married Relationship	18 (7.0)	14 (5.9)	12 (4.6)	44 (5.8)	8 (5.2)	2 (20.0)	0	10 (5.9)		54 (5.8)
	Long-Distance Romantic Relationship	5 (1.9)	1 (0.4)	0	6 (0.7)	1 (0.6)	0	0	1 (0.6)		7 (0.8)
Care	Childcare	148 (57.4)	168 (71.2)	124 (47.1)	444 (58.2)	89 (41.3)	5 (50.0)	3 (60.0)	97 (57.1)		541 (58.0)
	Eldercare	23 (8.9)	24 (10.2)	32 (12.2)	79 (10.4)	14 (9.0)	3 (30.0)	0	17 (10.0)		96 (10.3)
	Sandwiched Care	10 (3.9)	12 (5.1)	9 (3.4)	31 (3.9)	5 (3.2)	2 (20.0)	0	7 (4.1)		38 (4.1)
	Long-Distance Care	39 (15.1)	36 (15.3)	60 (22.8)	136 (17.8)	28 (18.1)	3 (30.0)	1 (20.0)	3 (18.8)		139 (14.9)

and postdocs (2.9 percent). Three-fourths were white (76.3 percent), and most (82.4 percent) were married or living with a romantic partner. A little less than one-tenth (7.7 percent) of married women faculty lived apart from their spouse or one of the spouses lived far from work because of the other's work. More than half (57.1 percent) provided care for children under the age of 18, 10 percent provided eldercare, and 7 percent provided sandwiched care (i.e., both childcare and eldercare). Nearly one-fifth, or 18.8 percent, provided care for family members who do not live with them. While most of the concerns of non-tenure-track faculty mirrored those of tenured and tenure-track faculty, we did notice some unique challenges, which we focus on here.

Most of the survey responses were qualitative and were analyzed using a content analysis method developed by Schreier (2012). First, we created our main coding frame, challenges and coping strategies, for each topic (e.g., childcare, eldercare, boundary management, work and nonwork, and effects) based on the literature review. Next, subcategories were created under each main category. They were defined to make sure each category was mutually exclusive and were continuously reexamined through discussion. After the coding was completed, we obtained final counts for each category. The full list of survey topics is provided in Table A-3.

The several quantitative items from the survey were analyzed using IBM SPSS Statistics 26 (IBM Corp., 2019). Means and standard deviations were obtained for the boundary-control measure to assess changes in boundary control. Using paired t tests, we also compared changes in prepandemic and postpandemic location preferences assessing the number of preferred and actual days working on and off campus over a 5-day week. For example, paired t-tests results revealed that, across the sample of women STEMM faculty, all reported significantly lower levels of boundary control after the pandemic than before the pandemic ($t = 33.42, p < .001$; 3.98 and 2.33 respectively). In order to examine the impacts of care responsibilities on the changes in the numbers of days working at home and boundary management, we used a general linear mixed model (Cnaan, et al., 1997; Krueger and Tian, 2004) approach. For example, a general linear mixed-model analysis was conducted to examine whether the magnitude of the increase in the number of days working at home post-COVID-19 was different between faculty with and without children. The result revealed that the increase in the number of days working at home post-COVID-19 pandemic was significantly greater for faculty with children than faculty without children ($F (1, 753) = 11.58$).

Literature Review Search Terms

To inform the literature review summarized in the commissioned paper, "Boundaryless Work: The Impact of COVID-19 on Work-Life Boundary Management, Integration, and Gendered Divisions of Labor for Academic Women in STEMM," We used the search criteria and obtained the number of results presented in Table A-4. To note, we found that, at the time of writing, few empirical papers focused specifically on COVID-19 and women in STEMM.

TABLE A-3 Topics for the October 2020 Survey

Quantitative Questions

Work location

| The number of days of working at home (out of 3 working days) before and after the pandemic | The preferred number of days of working at home |

Boundary control

The actual and preferred levels of boundary control before and after the pandemic, using a boundary control measure by Kossek et al. (2012)

Qualitative Questions

The positive and negative impact of COVID-19 on personal and career well-being

Boundary management

| The challenges of boundary management between work and family due to COVID-19 | Examples of boundary-setting practices (physical, temporal, mental, and technological) | Differences in boundary management between work and family after the pandemic |

Support from the university

| Examples of university support for work-life integration | What needs to be improved |

Housework demands

| The positive and negative impact of COVID-19 on nonwork responsibilities | The negotiation of nonwork responsibilities during the pandemic |

Care Demands

The positive and negative impact of COVID-19 on the following:

| Childcare | Eldercare | Sandwiched care | Long-distance care | Long-distance romantic relationship |

Background Information

| Gender | Race/ethnicity | Academic affiliation (university / department) | Tenure status (tenured, tenure-track but not yet tenured, non-tenure-track) | Rank |

TABLE A-4 Search Terms and Numbers of Results for the Literature Review Conducted by Kossek, Dumas, and Allen

Kossek Search	Psychinfo	Academic Search Complete
1. COVID-19 and Leadership	51	367
2. COVID-19 and HR	6	72
3. COVID-19 and organizational support	2	3
4. COVID-19 and faculty	148	21
5. COVID-19 and women faculty	2	4
6. COVID-19 and university	1,175	108
7. COVID-19 and higher education	22	24
8. COVID-19 and professors	11	0
9. COVID-19 and coping	108	1
10. COVID-19 and faculty stress	1	0
11. COVID-19 and faculty well-being	0	0
12. COVID-19 and faculty coping	0	0
13. COVID-19 and faculty eldercare	0	0
14. COVID-19 and faculty childcare	2	0
15. COVID-19 and faculty parenting	0	0
16. COVID-19 and faculty sandwiched care	0	0
17. COVID-19 and eldercare	5	0
18. COVID-19 and childcare	20	21
19. COVID-19 and STEM	10	122
20. COVID-19 and STEMM	0	0
Subtotal	1,563	743
Chronicle of Higher Education COVID 19 used as search term	769	
Total (Kossek)		3,075
Dumas Search		
1. "COVID-19" "women" "faculty' "stem"		3,290
2. "COVID-19" "women" "faculty" "stem" "U.S."		2,890
Allen Search		
USF library		
1. COVID-19 AND academic women		1,149
2. COVID-19 AND academic women AND division of labor		108
3. COVID-19 AND STEM		138,794
4. COVID-19 AND STEMM		105
SocArXIC Papers website		
1. COVID academic women		74

Appendix B

Methodology and Data Sources for the "Academic STEMM Labor Market, Productivity, and Institutional Responses," by Felicia A. Jefferson, Matthew T. Hora, Sabrina L. Pickens, and Hal Salzman

To understand the COVID-19 pandemic's potential effects on women in science, technology, engineering, mathematics, and medicine (STEMM), the authors of the paper "Academic STEMM Labor Market, Productivity, and Institutional Responses" that provided much of the information for Chapter 3, compiled two bodies of evidence. First, existing empirical literature concerning women in STEMM informed the pre-COVID-19 pandemic background. This literature is situated in many academic disciplines and addresses gendered and racialized barriers, among others, that are unique to STEMM (e.g., field or laboratory-related work pressures) and some that are relevant to the academic workforce as a whole (e.g., the utility of tenure clock extensions). Second, recent stories and emerging studies provide understanding of the COVID-19 pandemic's emerging effects. This work was conducted in fall 2020, within the first 9 months of the COVID-19 pandemic, which began in winter 2019. Given this timeline, much of the evidence used to assess how women academics are experiencing the COVID-19 pandemic and how it is affecting their careers is emergent. Additionally, it is important to stress that evidence of the impacts of the pandemic on some topics for particular groups (e.g., views of academic productivity for Black, Indigenous, and other Scholars of Color) were not yet available and thus are not included in Chapter 3 or throughout the report. In other cases, such as for academic productivity for women, considerable evidence exists but does not account for racial and ethnic diversity.

To identify the literature, the authors of this paper conducted a search of Google Scholar and PubMed using combinations of the following search terms: "STEMM," "careers," "pandemic," "intersectionality," and "women." The authors then reviewed the papers and conference presentations identified and further limited the search using the following criteria: the manuscripts had to be in English;

161

the materials had to be peer reviewed; and the content of the materials had to address the impacts of the COVID-19 pandemic on STEMM careers, the unique or disproportionate impact of the pandemic on women and/or Black, Indigenous, and People of Color researchers in the STEMM disciplines, and institutional responses to the pandemic.

The authors then analyzed the materials using thematic analysis techniques, which in this case included taking notes about recurring topics, data points, or arguments in articles, papers, and reports. The topics, data points, and arguments that were most frequently noted and/or were considered salient to the topic of this report were included in the submitted paper (Ryan and Bernard, 2003). These results served as the raw material for preparing this paper, which is laid out in three parts related to the effects of the COVID-19 pandemic on (1) the STEMM landscape (or STEMM educational attainment and STEMM occupations), (2) notions of academic productivity, and (3) institutional responses. Each part provides information as available about researchers at different career stages; Scholars of Color; effects by gender, specifically on women; and discipline and institution type.

Appendix C

Material Selection Process for "The Impact of COVID-19 on Collaboration, Mentorship and Sponsorship, and Role of Networks and Professional Organizations," by Misty Heggeness and Rochelle Williams

Misty Heggeness and Rochelle Williams, the commissioned authors for "The Impact of COVID-19 on Collaboration, Mentorship and Sponsorship, and Role of Networks and Professional Organizations," conducted an extensive review of recent articles on the academy; women; science, technology, engineering, mathematics, and medicine (STEMM); the COVID-19 pandemic; collaborations; and networks both in scientific journals and in the broader public domain of media, newspapers, and blogs. In addition, they conducted a literature review for rigorous evaluations on the impact of mentoring, professional organizations, and institutions in enhancing professional development and advancement. The authors of this paper viewed webinars (live and recorded) by various societies and networks, including the ADVANCE Resource and Coordination Network, American Economics Association Committee on the Status of Women in the Economics Profession, and others to cull resources.

They also reviewed the websites of professional scientific, engineering, and medical organizations and networks, such as the American Geophysical Union, Society of Women Engineers, and the Aspire Alliance, to determine how those groups are responding to membership needs. The complete list is available in Table C-1. To collate the list of professional STEMM organizations, they first selected organizations represented by membership in the Council of Engineering and Scientific Society Executives. They also utilized the comprehensive list of professional organizations and associations complied on https://jobstars.com/science-professional-associations-organizations. They also did special searches for computing, mathematical, and medical professional organizations to ensure adequate representation.

TABLE C-1 Professional Associations Reviewed by Misty Heggeness and Rochelle Williams

AACE International (Association for the Advancement of Cost Engineering)	American Statistical Association	International Society for Magnetic Resonance in Medicine
ABET (Accreditation Board for Engineering and Technology), Inc.	American Vacuum Society	International Society for Pharmaceutical Engineering, Inc.
ABSA International	American Water Resources Association	International Society for Stem Cell Research
Academies Collaborative	American Water Works Association	International Society of Automation
Acoustical Society of America	American Welding Society	International Society of Explosives Engineers
AIChE (American Institute of Chemical Engineers)	AOAC International	International Solar Energy Society
AIP (American Institute of Physics)	ASHRAE (American Society of Heating, Refrigerating and Air-Conditioning Engineers)	International Urogynecological Association
Alliance of Crops, Soils & Environ. Scientific Societies	ASM International	Laser Institute of America
American Anthropological Association	ASQExcellence	Linguistic Society of America
American Association for Anatomy	Association for Behavioral Analysis International	Materials Research Society
American Association for Cancer Research, Inc.	Association for Computing Machinery	Mathematical Association of America
American Association for Clinical Chemistry (AACC)	Association for Facilities Engineering	Metal Powder Industries Federation
American Association for the Advancement of Science (AAAS)	Association for Hospital Medical Education	NACE International
American Association of Bioanalysts	Association for Information Science and Technology	National Academy of Engineering
American Association of Engineering Societies	Association for Iron and Steel Technology	National Association of Corrosion Engineers
American Association of Petroleum Geologists	Association for Molecular Pathology	National Association of Multicultural Engineering Program Advocates, Inc.
American Association of Pharmaceutical Scientists	Association for Psychological Science	National Council of Examiners for Engineering and Surveying

APPENDIX C

American Association of Physicists in Medicine	Association for Public Policy Analysis and Management (APPAM)	National Council of Structural Engineers Associations
American Association of Physics Teachers	Association for Research in Vision and Ophthalmology	National Council of Teachers of Mathematics
American Association of Textile Chemists and Colorists	Association for the Advancement of Artificial Intelligence	National Environmental Health Association
American Astronomical Society	Association for the Advancement of Cost Engineering	National Foundation for Infectious Diseases
American Ceramic Society	Association for the Advancement of Medical Instrumentation	National Institute of Building Sciences
American Chemical Society	Association for Women in Computing	National Society of Black Engineers
American College of Veterinary Internal Medicine	Association for Women in Mathematics	National Society of Professional Engineers
American Concrete Institute	Association for Women in Science	North American Association for Environmental Education
American Council of Engineering Companies	Association of American Medical Colleges	Oceanic Engineering Society
American Crystallographic Association	Association of American Veterinary Medicine Colleges (AAVMC)	OSA - The Optical Society
American Dairy Science Association	Association of Bimolecular Resource Facilities	oSTEM Inc.
American Ecological Engineering Society	Association of Clinical Research Professionals	Population Association of America
American Epilepsy Society	Association of Conservation Engineers	Psychonomic Society
American Geophysical Union	Association of Energy Engineers	SACNAS (Society for Advancement of Chicanos/Hispanics and Native Americans in Science)
American Indian Science and Engineering Society	Association of Environmental & Engineering Geologists	SAE International
American Industrial Hygiene Association	ASTM International	SAMPE (Society for the Advancement of Material and Process Engineering)
American Institute for Medical & Biological Engineering	Audio Engineering Society	Seismological Society of America
American Institute of Aeronautics & Astronautics	BICSI	Sexual Medicine Society of North America

APPENDIX C

American Institute of Architects	Biomedical Engineering Society	SHPE (Society of Hispanic Professional Engineers)
American Institute of Chemists	Biophysical Society	Society for Biological Engineers
American Institute of Mining, Metallurgical, and Petroleum Engineers, Inc.	Board of Certified Safety Professionals	Society for Conservation Biology
American Institute of Physics	Botanical Society of America	Society for Imaging Science & Technology
American Institute of Professional Geologists	Casualty Actuarial Society	Society for Industrial & Applied Mathematics
American Mathematical Association of Two-Year Colleges	Clinical Laboratory Management Association	Society for Industrial and Organizational Psychology
American Mathematical Society	Coalition for Academic Scientific Computation	Society for Industrial Microbiology and Biotechnology
American Medical Colleges	Coastal and Estuarine Research Federation	Society for Investigative Dermatology
American Meteorological Society	Computing Research Association	Society for Mining, Metallurgy, & Exploration, Inc
American National Standards Institute	Computing Science Accreditation Board (CSAB)	Society for Neuroscience
American Nuclear Society	Construction Management Association of America	Society for Pediatric Research
American Oil Chemists' Society	Construction Specifications Institute (CSI)	Society for Sedimentary Geology
American Ornithological Society	Council for Agricultural Science and Technology	Society for the Study of Reproduction
American Pediatric Society	Council of Engineering and Scientific Specialty Boards	Society of Actuaries
American Physical Society	Council of Landscape Architectural Registration Boards (CLARB)	Society of Allied Weight Engineers
American Physiological Society	Council of Scientific Society Presidents	Society of American Military Engineers
American Phytopathological Society	Council on Undergraduate Research	Society of Asian Scientists and Engineers
American Psychological Association	Directed Energy Professional Society	Society of Cable Telecommunication Engineers
American Public Works Association	Ecological Society of America	Society of Economic Geologists

APPENDIX C

American Society for Biochemistry and Molecular Biology	Electrochemical Society	Society of Environmental Toxicology & Chemistry
American Society for Cell Biology	Entomological Society of America	Society of Exploration Geophysicists
American Society for Clinical Pharmacology & Therapeutics	Environmental Engineering Geophysical Society	Society of Fire Protection Engineers
American Society for Engineering Education	Fabricators and Manufacturers Association, International	Society of Flight Test Engineers
American Society for Engineering Management	Federation of American Societies for Experimental Biology	Society of Manufacturing Engineers
American Society for Healthcare Engineering	Federation of Associations in Behavioral & Brain Sciences	Society of Mexican-American Engineers and Scientists
American Society for Microbiology	Federation of Materials Societies	Society of Motion Picture and Television Engineers
American Society for Nondestructive Testing, Inc.	Fluid Power Society	Society of Naval Architects and Marine Engineers
American Society for Parenteral and Enteral Nutrition	Geochemical Society	Society of Petroleum Engineers, Inc.
American Society for Pharmacology and Experimental Therapeutics	Geological Society of America, Inc.	Society of Plastics Engineers
American Society for Quality	GeoScienceWorld	Society of Reliability Engineers
American Society of Agricultural and Biological Engineers	Human Factors and Ergonomics Society	Society of Toxicology
American Society of Certified Engineering Technicians	IFSCC (International Federation of Societies of Cosmetic Chemists)	Society of Wetland Scientists
American Society of Civil Engineers	Illuminating Engineering Society of North America	Society of Women Engineers
American Society of Gas Engineers	IMAPS-International Microelectronics Assembly and Packaging Society	Soil and Water Conservation Society
American Society of Heating, Refrigerating and Air Conditioning Engineers	Industrial Research Institute	Southeastern Consortium for Minorities in Engineering
American Society of Human Genetics	Infectious Diseases Society of America	SPE - Inspiring Plastics Professionals
American Society of Landscape Architects	INFORMS (Institute for Operations Research and the Management Sciences)	SPIE (The International Society for Optical Engineering)

American Society of Materials, International	Institute of Biological Engineering	Standards Engineering Society
American Society of Mechanical Engineers	Institute of Electrical and Electronics Engineers, Inc.	TAPPI (Technical Association of the Pulp and Paper Industry)
American Society of Naval Engineers	Institute of Environmental Sciences and Technology	The American Ceramic Society, Inc.
American Society of Parasitologists	Institute of Food Technologists	The American Medical Women's Association
American Society of Plant Biologists	Institute of Industrial and Systems Engineers	The Ecological Society of America
American Society of Safety Engineers	Institute of Mathematical Statistics	The Electrochemical Society
American Society of Test Engineers	Institute of Transportation Engineers	The Endocrine Society
American Society of Tropical Medicine and Hygiene	International Association of Medical Science Educators (IAMSE)	The Histochemical Society
American Sociological Association	International Federation for Medical and Biological Engineering	The Minerals, Metals & Materials Society
American Speech-Language-Hearing Association	International Society for Computational Biology	Women in Engineering Programs and Advocates Network

Appendix D

Committee Biographies

DR. EVE HIGGINBOTHAM (NAM) is the inaugural vice dean for inclusion and diversity of the Perelman School of Medicine at the University of Pennsylvania, a position she assumed on August 1, 2013. She is also a senior fellow at the Leonard Davis Institute for Health Economics and professor of ophthalmology at the University of Pennsylvania. She has been a member of the National Academy of Medicine (NAM) since 2000 and is now an elected member of the NAM Council, upon which she chairs the Finance Committee. Dr. Higginbotham is also a member of the Governing Board of the National Research Council and past president of the American Optometric Association (AOA) Medical Honor Society. Notable prior leadership positions in academia include dean of the Morehouse School of Medicine, senior vice president for health sciences at Howard University, and professor and chair of the Department of Ophthalmology and Visual Sciences at the University of Maryland in Baltimore, a position she held for 12 years. She formerly chaired her section of the NAM and is a former member of the NAM membership committee. Dr. Higginbotham also serves as an associate editor on the Editorial Board of the *American Journal of Ophthalmology*. Dr. Higginbotham, a practicing glaucoma specialist at the University of Pennsylvania, has either authored or coauthored more than 150 peer-reviewed articles and coedited four ophthalmology textbooks. She continues to remain active in scholarship related to glaucoma, health policy, STEM, and patient care. She holds undergraduate and graduate degrees in chemical engineering from MIT, and a medical doctorate from Harvard Medical School, completed her residency in ophthalmology at the Louisiana State University Eye Center, and completed a master's of law degree from the University of Pennsylvania Carey Law School.

DR. ELENA FUENTES-AFFLICK (NAM) is a professor and vice chair of pediatrics and chief of pediatrics at the Zuckerberg San Francisco General Hospital, and vice dean for academic affairs in the School of Medicine at University of California, San Francisco (UCSF), a position she has held since 2012, She is responsible for overseeing all academic affairs in the School of Medicine, including the recruitment, development, and advancement of a diversified academic workforce of the highest caliber. She is also responsible for overseeing innovative programs for faculty orientation, career development, and leadership training. After completing her undergraduate and medical education at the University of Michigan, Dr. Fuentes-Afflick came to UCSF for her residency training in pediatrics, followed by a fellowship in health policy at UCSF and M.P.H. in epidemiology from UC Berkeley. She has served in several important national leadership roles, including president of the Society for Pediatric Research, president of the American Pediatric Society, and service on the Council of the National Institute for Child Health and Development. Dr. Fuentes-Afflick was chair of the UCSF Academic Senate from 2009 to 2011. She was elected to the National Academy of Medicine in 2010 and has served as a member of several consensus study committees.

DR. LESLIE D. GONZALES is an associate professor in the Higher, Adult, and Lifelong Learning unit at Michigan State University in the College of Education. She also serves as an affiliate faculty member in the Center for Gender in a Global Context and Chicano/Latinx studies. As a Latina, first-generation-college-student-turned-academic who earned all three of her academic degrees from Hispanic-Serving Institutions, Dr. Gonzales now studies how relations of power, privilege, and prestige operate in ways that can be detrimental to historically underrepresented scholars. In addition to her faculty role, Dr. Gonzales serves as the faculty advocate for the College of Education at Michigan State University—a role that allows her to apply her research to practice and advance equity-oriented practices and systems change. Dr. Gonzales is currently the co-principal investigator on Aspire, a multimillion dollar project sponsored by the National Science Foundation.

DR. JENI HART is the dean of the Graduate School and vice provost for graduate studies at the University of Missouri. She is also professor of higher education in the Department of Educational Leadership and Policy Analysis (ELPA). Dr. Hart joined ELPA as an assistant professor in 2003. She completed her Ph.D. in higher education administration at the University of Arizona. Prior to becoming a faculty member, she worked for 9 years as a student affairs educator at a number of colleges and universities, and 1 year as a faculty member at Southeast Missouri State University. Dr. Hart's scholarship centers on three mutually reinforcing themes: faculty work, gender and feminisms, and campus climate. Specifically, she is interested in how organizational structures in academe mutually shape the experiences of those in higher education, particularly women and feminist faculty. Dr. Hart serves on the editorial boards of the *Journal of Diversity in Higher Education* and the *NASPA Journal about Women in Higher Education*.

DR. RESHMA JAGSI, M.D., D.Phil., is Newman Family Professor and deputy chair in the Department of Radiation Oncology and director of the Center for Bioethics and Social Sciences in Medicine at the University of Michigan. She graduated first in her class from Harvard College and then pursued her medical training at Harvard Medical School. She also served as a fellow in the Center for Ethics at Harvard University and completed her doctorate in social policy at Oxford University as a Marshall Scholar. A substantial focus of her research considers issues of bioethics and gender equity in academic medicine. Her investigations of women's underrepresentation in senior positions in academic medicine and the mechanisms that must be targeted to promote equity have been funded by two National Institutes of Health R01 grants and grants from the Robert Wood Johnson Foundation, American Medical Association, and other philanthropic funders. She leads the national program evaluation for the Doris Duke Charitable Foundation's Fund to Retain Clinician Scientists, a large national intervention that was inspired in part by her own research. Active in organized medicine, she has served on the Steering Committee of the American Association of Medical Colleges' Group on Women in Medicine in Science and now serves on the Board of Directors of the American Society of Clinical Oncology. She was part of the *Lancet*'s advisory committee for its theme issue on women in science, medicine, and global health.

DR. LEAH JAMIESON (NAE) is Ransburg Distinguished Professor of Electrical and Computer Engineering at Purdue University and John A. Edwardson Dean Emerita of the College of Engineering, and holds a courtesy appointment in Purdue's School of Engineering Education. She is a member of the National Academy of Engineering (NAE) and the American Academy of Arts and Sciences and a fellow of the American Society for Engineering Education and the Institute of Electrical and Electronics Engineers (IEEE). She is cofounder and past director of the EPICS (Engineering Projects in Community Service) program. She was an inaugural recipient of the National Science Foundation Director's Award for Distinguished Teaching Scholars. Dr. Jamieson served on the steering committee for the NAE report *Changing the Conversation: Developing Effective Messages for Improving Public Understanding of Engineering*, the National Research Council report *Barriers and Opportunities for 2-Year and 4-Year STEM Degrees: Systemic Change to Support Students' Diverse Pathways*, and the National Academies report *Cultivating Interest and Competencies in Computing: Authentic Experiences and Design Factors*. She has served as president and chief executive officer of the IEEE, board chair of the Anita Borg Institute, and cochair of the Computing Research Association's Committee on the Status of Women in Computing Research. She received an S.B. in mathematics from MIT and M.A., M.S.E, and

Ph.D. degrees in electrical engineering and computer science, all from Princeton University. She has been awarded honorary doctorates by Drexel University and the New Jersey Institute of Technology.

DR. ERICK C. JONES is a professor and associate dean for graduate studies in the College of Engineering at the University of Texas at Arlington (UTA). He is currently the George and Elizabeth Pickett Endowed Professor in Industrial, Manufacturing and Systems Engineering. Dr. Jones returned from his 3-year rotating detail at the National Science Foundation (NSF), where he was a program director in the Engineering Directorate for Engineering Research Centers Program. Earlier Dr. Jones worked at NSF in the Education Directorate, where he worked in the Division of Graduate Education and led the INTERN and Graduate Research Internship Programs. He was also a program director for the prestigious Graduate Research Fellowship Program. Dr. Jones was one of the few program officers who worked in two directorates as a rotating program director. Prior to joining UTA, Dr. Jones worked at the University of Nebraska-Lincoln for 8 years, where he initially received tenure. He served as deputy director of UTA's Security Advances via Nanotechnologies Center from 2013 to 2015. He is an active member of American Association for the Advancement of Science, Institute of Industrial and Systems Engineers (IISE), American Society for Engineering Education, and National Society of Black Engineers (NSBE). Dr. Jones has served IISE, NSBE, and other organizations as faculty advisor for the past decade; served as an Alfred Sloan Minority Ph.D. program director and now on the Sloan Mentoring Network Board; has worked with the National Action Council for Minorities in Engineering for more than a decade; and was one of the initial founders and past chair of Texas A&M's Black Former Students Network. Dr. Jones was recognized as an Alfred Sloan Underrepresented Minority Ph.D. Program Fellow and has been honored by the National Action Council for Minorities in Engineering three times. Dr. Jones earned his bachelor's degree in industrial engineering from Texas A&M University and his master's and doctoral degrees in industrial engineering from the University of Houston.

DR. BERONDA MONTGOMERY is professor of biochemistry and molecular biology and microbiology and molecular genetics in the Department of Energy Plant Research Laboratory at Michigan State University (MSU). She completed doctoral studies in plant biology at the University of California, Davis, and was a National Science Foundation (NSF)–funded postdoctoral fellow in microbial biology at Indiana University. Since starting at MSU in 2004, Dr. Montgomery's laboratory investigates the mechanisms by which organisms such as plants and cyanobacteria that have limited mobility are able to monitor and adjust to changes in their external environment. The ability of these largely immobile organisms to adapt their patterns of growth and development to fluctuations in external environmental parameters increases their survival and maximizes their growth and productivity. Dr. Montgomery's scholarly efforts were recognized by her receipt of an NSF CAREER Award in 2007, being selected as a finalist in the 2014 Howard Hughes Medical Institute Professors Competition, and a 2015 Michigan State University Nominee for the Council for Advancement and Support of Education U.S. Professor of the Year Award. In addition to her core research and teaching efforts, Dr. Montgomery is also actively involved in scholarly efforts to promote effective research mentoring and management and the inclusion and success of individuals from groups underrepresented in the sciences. She has published extensively on evidence-based strategies to nurture and retain talent in academia, developing strategies for effective mentorship that center on the individual and their specific needs and goals. As an expert in effective and evidence-based mentorship, Dr. Montgomery serves on a number of leadership boards and as a consultant to universities working toward greater diversity, equity, and inclusion within their research and education programs.

DR. KYLE MYERS is an assistant professor of business administration in the Technology and Operations Management unit of Harvard Business School (HBS). Dr. Myers studies the economics of innovation. His research lies at the intersections of science, health care, and the commercialization process. More specifically, Professor Myers is interested in the strategic choices and performance of scientists, the supply and demand of innovation in high-tech sectors, public versus private funding of research and development, and the management of innovation in large organizations such as hospitals and pharmaceutical and engineering firms. His work has received funding from the Kauffman Foundation and was awarded the National Bureau of Economic Research - Institute of Fiscal Studies Predoctoral Scholarship in the Value of Medical Research. Professor Myers holds a Ph.D. from the Wharton School's Department of Health Care Management and Economics. He has an M.S. in health policy and management and a B.S. in biology from Penn State University. Prior to joining HBS, he served as a postdoctoral fellow at the National Bureau of Economic Research and worked at the Centers for Disease Control and Prevention.

DR. RENETTA TULL is the vice chancellor of diversity, equity and inclusion at the University of California, Davis. Before joining UC Davis in 2019, Dr. Tull was associate vice provost for strategic initiatives at the University of Maryland, Baltimore County (UMBC), and professor of the practice in UMBC's College of Engineering and Information Technology (COEIT). Within COEIT, she served as part of the "Engagement" team and pursues research in humanitarian engineering. Dr. Tull is founding director and co-principal investigator for the 12-institution National Science Foundation (NSF) University System of Maryland's (USM) PROMISE Alliances for Graduate Education and the Professoriate, and codirector/co-principal investigator for the NSF USM's Louis Stokes Alliance for Minority Participation. In addition to roles at UMBC and roles with grants, she also served USM as special assistant to the senior vice chancellor for academic affairs and student affairs, and was the system's director of graduate and professional pipeline development. In 2017, Dr. Tull was appointed to serve as chair for USM's Health Care Workforce Diversity subgroup. Dr. Tull has engineering and science degrees from Howard University and Northwestern University. Dr. Tull served on the Science of Effective Mentorship in STEMM consensus study committee.

MS. MARIA LUND DAHLBERG is a senior program officer and study director with the Board on Higher Education and Workforce and the Committee on Women in Science, Engineering, and Medicine of the National Academies of Sciences, Engineering, and Medicine. Her current work focuses on the Impact of COVID-19 on the Research Careers of Women in Academic Science, Engineering, and Medicine; the Response and Adaptation of Higher Education to the COVID-19 Pandemic; the Science on Effective Mentoring in STEMM (Science, Technology, Engineering, Mathematics, and Medicine); and Equity, Diversity, and Inclusion in Postsecondary Education. Her work with the National Academies spans topics ranging from equity and identity in science, through science communications, to postdoctoral research experiences, health care, and innovation ecosystems. She came to the National Academies by way of a Christine Mirzayan Science and Technology Policy Fellowship, which she received after completing all requirements short of finalizing the dissertation for her doctorate in physics at the Pennsylvania State University. Ms. Lund Dahlberg holds a B.A. with high honors in physics from Vassar College and an M.S. in physics from Pennsylvania State University.